INDIVIDUALITY AND DETERMINISM
Chemical and Biological Bases

INDIVIDUALITY AND DETERMINISM
Chemical and Biological Bases

Edited by
Sidney W. Fox
University of Miami
Coral Gables, Florida

PLENUM PRESS • NEW YORK AND LONDON

Library of Congress Cataloging in Publication Data

Liberty Fund Conference on Individuality and Determinism (1982: Key Biscayne, Fla.)
    Individuality and determinism.

    "Proceedings of the Liberty Fund Conference on Chemical and Biological Bases for In-
dividuality, held May 1–2, 1982, at Key Biscayne, Florida"—T.p. verso.
    Includes bibliographical references and index.
    1. Molecular biology—Philosophy—Congresses. 2. Evolution—Philosophy—Congresses.
3. Individuality—Congresses. 4. Free will and determinism—Congresses. 5. Nature and nurture—Congresses. I. Fox, Sidney W. II. Liberty Fund. III. Title.
QH506.L45   1982   574'.01   84-2103
ISBN-13: 978-1-4615-9381-2   e-ISBN-13: 978-1-4615-9379-9
DOI: 10.1007/978-1-4615-9379-9

Proceedings of the Liberty Fund Conference on Chemical and Biological Bases for
Individuality, held May 1–2, 1982 at Key Biscayne, Florida

©1984 Plenum Press, New York
Softcover reprint of the hardcover 1st edition   1984
A Division of Plenum Publishing Corporation
233 Spring Street, New York, N.Y. 10013

PREFACE

The significance of human individuality is such that each human functions as a unique "molecular" unit of the mass of humanity. Understanding the natural basis for the uniqueness of the individual has long been an objective. The possibilities have been analyzed by Julian Huxley, by A. E. Needham, by Roger Williams, and by others.

With his books Biochemical Individuality and Free and Unequal, Roger Williams has done as much as anyone to focus attention on this complex of questions. Although scheduled to participate in this program, Roger Williams* was unable to attend due to illness. He asked, however, that a quotation be included in the proceedings. This quotation from Chraka is presented early in this book.

While metabolic bases for individuality have received a considerable investigation and discussion by Williams and others, the case for underlying determinants and derivative consequences have not been examined as fully. The specificities that abound in our living world can be traced to the manner in which molecules fit with each other. While numerous studies having other objectives can be cited in support of molecularly based specificities, a few of the leaders in the development of the understanding of physical aspects of biological information present here some of their latest inferences. Several of the participants discuss some of the consequences at higher levels. Examination of the fascinating cases of reunited identical twins are seen as providing a capstone to the hierarchical treatment.

When asked to consider the possibility of organizing a first group of basic scientists to determine how they might interact in a conference on some subject of interest to the Liberty Fund, I saw this as a shiningly attractive opportunity. This opportunity was not the first one of its kind, even in the large leap from

---

*As the dean of biochemical individuality, Roger Williams was honored by the conference. This was unfortunately done in absentia.

molecules to society. Not long before this conference I was
involved as a participant in the First Biennial T. C. Schneirla
Conference on Levels of Integration and Evolution of Behavior
(organized by the Department of Psychology of Wichita State University and others). At that conference, with its center of gravity
in comparative psychology, it was possible to respond to a question
of the origin of sociality posed by Tobach and Schneirla in 1968.
The full hierarchical treatment for behavior, emerging from rudimentary molecular interactions, can be found in the book titled
Levels of Integration and Evolution of Behavior, edited by
G. Greenberg and E. Tobach.

The Liberty Fund Conference on Key Biscayne had as its theme
the underlying chemical and biological bases for individuality, a
subject of obvious relevance to those wishing to understand liberty
and democratic institutions.

While the authors were each free to develop his or her topic
in his or her own way, as is customary, the unravelling of the
conference was remarkable for its emphasis on determinism from Koshland to Bouchard, and originating with Williams' quotation of
Chraka. This emphasis on the individual and on determinism seems
to be part of a developing outlook on genetic determinism that spans
from T. H. Morgan to E. O. Wilson, and that now reflects a growing
understanding of molecular determinism. The latter is derived from
the biochemical specificity of enzyme-substrate interactions, from
genetic coding, and from the internally directed pathways of molecular evolution. The striking nature of genetic determinism, which
has slowly but inexorably captured the mind of man, is increasingly
seen to be rooted in the "chemical and physical composition and
configuration" of living things of which Morgan wrote. Bouchard's
chapter on reunited twins obviously stresses determinism, as does
Koshland's chapter in which he extrapolates from bacterial behavior
to determinism and analyzes its relationship to environment and
random fluctuations. Other remarks in the discussion sections
represent various shades of agreement with, or disagreement with,
the deterministic hypothesis. Mention of a recent volume titled
Against Biological Determinism, edited by Steven Rose (Allison and
Busby), is broadening in this context, as is also Popper's, The
Open Universe: An Argument for Indeterminism.

Not all of the papers in this book were presented at the conference. Since two of the original eight papers were not at first
available, due to unfortunate and inconvenient hospitalization of
the authors, the book was enlarged by one paper by a physicist,
Koichiro Matsuno, visiting professor at this Institute, and by a
chapter by the organizer.

We thank the Liberty Fund staff for their friendly instigation
and for responding to questions of policy, mostly by skillful neglect.

Maynard Dockendorf, as administrative assistant in this Institute, saw to it that the organization of the meeting ran smoothly. Dr. Charles Metz, research professor and participant, provided much helpful advice.

Peggy Nemeth supervised the collection of manuscripts and related chores. Grace Matheson and Jennie Myers aided in the typing.

Sidney Fox

"The individual constitution (Prakriti) is an inherited condition that cannot be altered fundamentally. It is a life-long concern for every individual. This factor of individual personality is of supreme significance in determining the conditions of health and disease in man."

By Chraka (about 3000 B.C. from the Sanskrit)

CONTENTS

INDIVIDUALITY IN BACTERIA AND ITS RELATIONSHIP TO HIGHER SPECIES

Daniel E. Koshland, Jr.

University of California
Berkeley, California 94720

INTRODUCTION

Individuality is at once one of the most highly prized posses-
sions of human beings, and at the same time the source of some of
our greatest difficulties. Each of us wants to be an individual
and, in fact, a great preoccupation of the current press and psychiatric social workers is to be devoted to "finding one's identity." Even the news media recognizes, however, that we inherit
a certain amount of our behavior, and that many of our characteristics are common to our fellowman of quite different heredities and
who have grown up in quite different environments. One of the
important assessments of human behavior, therefore, becomes an
attempt to identify those features that are individual and those
that are characteristic of the species or of some large subset of
the species.

We are beginning to become aware that drug damages are not
common to all individuals of the same height and weight, and in
fact vary distinctly from one individual to the next. Susceptibility to diseases such as cancer and multiple sclerosis appears
to be different from individual to individual, based to a considerable extent on his or her genetic makeup. It becomes important,
therefore, to examine the basis for individuality at the molecular
level to ascertain those features which would be probably or
inevitably common to all members of the species and others which
will vary from individual to individual.

As one travels up the evolutionary ladder, the species become
more similar to man and hence in some ways more interesting, but
inevitably become more difficult in terms of performing definitive

1

experiments. Simpler species are far more easy to control and,
therefore, more decisive experiments can be performed to separate
the properties of individuality and resolve the controversies of
nature and nurture. It will be asked whether simple species can,
in any way, be assumed to have any relationship to their more complex analogs. The answer which we will discuss below would seem
to be a resounding "yes," provided the analogies are restricted to
those features in which direct relationships can be identified.
For example, the bacterium is a single cell organism, having a distinct behavioral pattern which allows it to survive in a complex
and dangerous world. The processing system within a bacterium is
very similar to the processing system of individual cells in higher
species. Since bacteria are monocellular species, the interactions
of cells with each other, as in a neural network, becomes a more
dubious extrapolation, but in this case we are helped by other areas
of investigation which delineate the role of a single cell in the
complex matrix. Many of the advances of modern psychological understanding, for example, have occurred because of the tracing of
mental diseases to the actions of individual neurons, such as
deficiencies in dopamine neurotransmitters in Parkinsonism and
schizophrenia. To the extent that the bacterium gives us insight
into the alteration in the properties of individual cells, it may
be extrapolated to human individuality.

In the pages which follow, we shall describe various studies
in our laboratory in regard to the individuality of bacteria and
the influence of this individuality on the bacterial processing
system. We will then make some tentative suggestions of how this
could affect the behavior of more complex species.

Response Regulator Model for Chemotaxis

The ability of bacteria to control their migration has been
known since the 1880s, when the phenomenon was uncovered by Engelmann (1881) and Pfeffer (1883). Molecular processes which control
its behavior have been studied in recent times in a number of laboratories, of which Adler (1969) deserves special mention for reopen-
ing the area in modern times, and in which many laboratories,
including those of Simon, Parkinson, Macnab, Berg, Ordal, Hazelbauer,
Taylor, Dahlquist, and our own have made significant contributions.
Bacteria migrate by regulating tumbling frequency (Berg and Brown,
1972; Macnab and Koshland, 1972). Bacteria suppress tumbling when
they go in a favorable direction and increase tumbling when they go
in an unfavorable direction. A favorable direction reflects an
an unfavorable direction. A favorable direction reflects an
increasing gradient of attractant, which is usually a nutrient, or
decreasing gradient of repellent, which is usually an indicator of
toxic conditions. Unfavorable directions generate the opposite
response. The wide range of receptors available in a bacterium

(Adler, 1969; Hazelbaur and Parkinson, 1977) allows it to use this migration aptitude as a general regulatory device to optimize its environmental conditions. It is apparent that this ability to seek "pleasure" or avoid "pain" has strong analogies in the pain/pleasure surveillance system of higher mammalian species (Koshland, 1977).

It was also found that this control of tumbling frequency was achieved by a temporal sensing mechanism; i.e., bacteria have a rudimentary memory that allows them to analyze differences in concentration within a given time frame (Macnab and Koshland, 1972). A bacterium moving randomly through solution does not change behavior if the distribution of chemicals is isotropic. Under this circumstance, it swims in roughly straight lines and tumbles in random directions at a constant frequency (Berg and Brown, 1974). However, the same bacterium integrates stimuli in such a way as it moves through a medium that it will know when the concentration at t_2 is different from the concentration of t_1 (Macnab and Koshland, 1972). Then a signal is sent back through the processing system to the flagella motors, either generating or suppressing tumbling.

The sensing behavior can be rationalized by using the model shown in Figure 1 in which a response regulator is assumed to behave like a thermostat. When the response regulator concentration is above the threshold, tumbling is suppressed; below the threshold, tumbling is stimulated. Random variation around the threshold would lead to the random runs and tumbles observed in the absence of a gradient. A stimulus such as an increase in attractant concentration leads to an increase in response regulator concentration and hence to suppression of tumbling. A decrease in attractant leads to a decrease in response regulator concentration and hence to generation of tumbling. When stimulation is complete, the system reverts to normal; i.e., it adapts, even though the total level of chemoeffector has risen.

This tumble regulator model (Koshland, 1977; Macnab and Koshland, 1972) has led to many predictions that have been verified even though the model remains conceptual. Such a model would indicate that the gradients of repellents and attractants should have similar effects but opposite algebraic signs. This was indeed observed (Tsang et al., 1973; Tso and Adler, 1974). Furthermore, constantly tumbling mutants would have a tumble regulator below threshold but could be momentarily elevated by a strong gradient of attractant, as is observed (Aswad and Koshland, 1974). Finally, constantly smooth swimming mutants would have a gradient of attractant above threshold which could be temporarily depressed by a strong gradient of repellent, and this too occurred (Springer and Koshland, 1977).

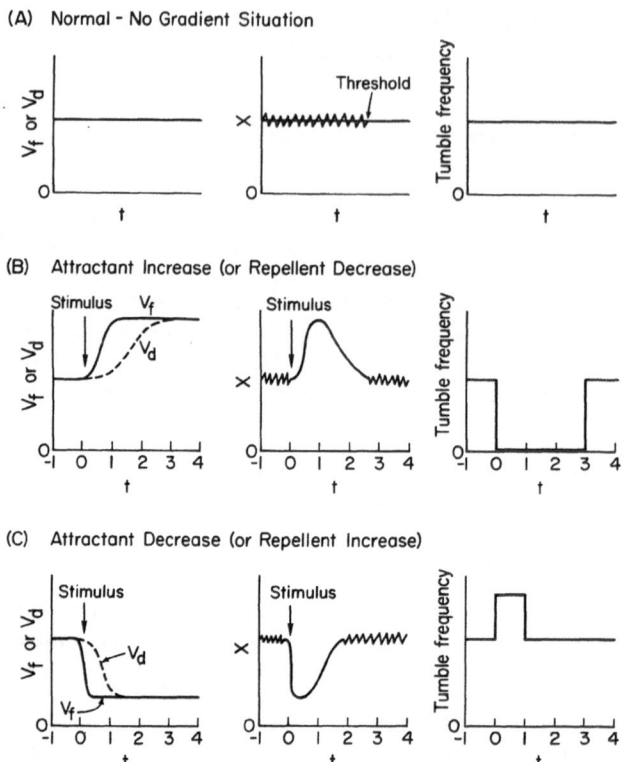

Fig. 1. Response of wild-type bacterium to attractants and repellents as explained by a response (tumble) regulator model. The
variation over time for the enzyme activities, the level of tumble
regulator and the tumbling frequency is shown for three situations.
(A) In absence of a gradient, $V_f=V_d$ are constant over time, and X
(the tumble regulator) concentration varies around threshold in a
Poissonian manner. The tumble frequency remains essentially constant. (B) Sudden increase in attractant increases rate of V_f
faster than V_d leading to transient increase in concentration of X
and transient decrease in tumbling frequency. Repellent decrease
gives the same effect. (C) Sudden decrease in repellent decreases
rate of V_f more rapidly than V_d leading to decrease in concentration of X and transient increase in tumbling frequency.

Chemotaxis as a Tool to Distinguish Mechanisms

Because the tools exist for analyzing chemotaxis in individual
bacteria and also because such analyses take very little time relative to the lifetime of a bacterium, chemotaxis seemed an ideal
vehicle with which to examine the relationships of heredity,
environment, and chance.

Other studies' have been made on bacterial variation (Delbrück, 1945; Powell, 1959; Stocker, 1949; Lederberg and Iino, 1956; Novick and Weiner, 1957; Maloney and Rotman, 1973; Benzer, 1953; Quadling and Stocker, 1957, 1962; Iino, 1975; Spiegelman, 1951; Cairns, 1960; Spudich and Koshland, 1976). Some variation can be explained by differences in the environment of individual cells. A colony on a nutrient plate can vary since the outside cells are not in the same environment as those inside. Variation in bacteria can also be ascribed to genetic mutation rate, environmental fluctuation, asynchrony of the population, and probabilistic considerations. The Spudich assays for chemotaxis and the ability to grow bacteria in swirled culture provided tools for distinguishing among these factors (Spudich and Koshland, 1976).

A bacterial clone arising from a single cell was subjected to a temporal gradient, and the adaptation phenomenon; i.e., the return of the bacteria to their normal tumbling state, was followed both for the population and for individual bacteria (Spudich and Koshland, 1976). A video tape could be played back in slow motion to study in detail the bacteria's behavior over a long period of time. Twenty-two individuals of a constantly tumbling mutant were stimulated with serine, then tethered to the wall of a cover glass following the method of Silverman and Simon (1974). The same strains of bacteria swimming freely in large colonies were studied by the photoassay technique (Spudich and Koshland, 1975), and the results are shown in Figure 2. Not only were the curves of response essentially identical for the colony and the individual, but the variation was Gaussian and far higher than one could explain on the basis of mutation rate.

If we define $F(t)$ as the fraction of a population that has recovered at time t and $S(t)$ as the probability of producing a smooth swimming track, equation (1) described the behavior of the bacteria. S_{oo} is 0 for the constantly tumbling mutant and has a finite value greater than 0 for wild type bacteria.

$$1 - F(t) = \frac{S(t) - S_{oo}}{1 - S_{oo}} \qquad (1)$$

The bacteria used in these experiments were, at most, thirty generations away from a single cell parent. An extremely high spontaneous mutation rate would have to be postulated to explain our data in terms of genetic variation, as the following calculation shows. Where m is the probability of occurrence of any mutation affecting chemotaxis per generation, after thirty generations the probability of a bacterium's not containing a mutation is $(1-m)^{30}$. Therefore, for at least 50% of the bacteria to have acquired one or more mutations affecting chemotaxis,

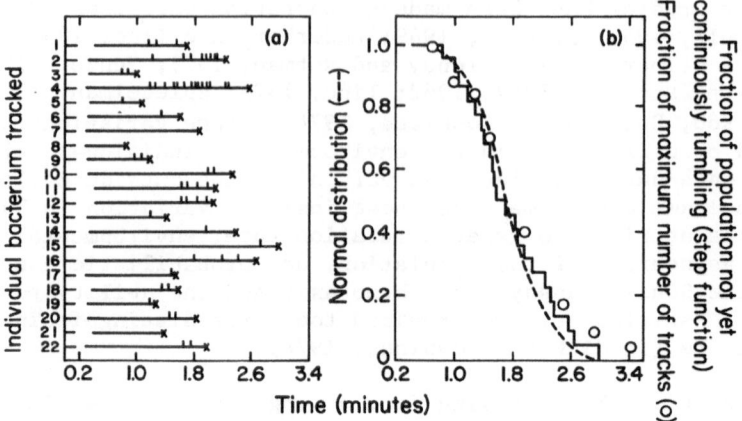

Fig. 2. Tumbly mutant recovery individuality. All bacteria were grown in homogeneous conditions as described previously (Tsang et al., 1973). In a the length of each line represents the time interval over which an individual bacterium was tracked. For each of the 22 lines the temporal gradient stimulus 0-0.5 mM L-serine was delivered to ST171 at time 0 (as previously described, Tsang et al., 1973) and the bacterium nearest the center was followed until it tumbled continuously for 30 s. Times at which the bacterium briefly tumbles are represented, and the time at which it began to continuously tumble is represented. From the data in (a) the fraction of the bacteria not yet continuously tumbling was plotted against time to yield the step function in (b). Tumble frequency assay (Tsang et al., 1973) results for ST171 subjected to the same stimulus,---. A cumulative normal distribution with the same mean and variance as the data obtained from tracking is represented.

$$(1-m)^{30} < 0.5$$

or

$$m > 2.3 \times 10^{-2}$$

Only about ten genes are specifically involved in Salmonella chemotaxis, but even a liberal estimate that a mutation in any 1 of 100 loci could alter chemotactic behavior requires that a mutation rate per locus must be greater than 2×10^{-4}, which is at least 10^{3} times greater than usual spontaneous mutation rates. Therefore, we conclude that (1) there is a large nongenetic individual variability in the response of both tumbling mutants and wild type bacteria to stimuli, and (2) the tumble frequency assay directly measures this variability.

Effect of Phase of the Cell Cycle

Two alternatives could explain the above results: (1) the
individuality is created by one or more random events in the history
of the bacterium, or (2) the cell has no intrinsic individuality, but
its properties vary according to its position in the cell division
cycle. In the latter case the asynchrony of the culture would give
the appearance of individuality when it did not, in fact, exist.
To distinguish between these alternatives, the responses of the bacteria were studied as a function of cell length. Cell length is
known to correlate with position in the cell division cycle (Schaecter
et al., 1962). The lengths of smooth swimming tumbly mutant bacteria
after stimulation at a time when 95% of the population was smooth
swimming were compared with those when 10% were still smooth swimming.
The former gives values for the overall population and the latter
for the longest responding 10% of the bacteria. The distributions
of the bacterial lengths are shown in Table 1. There is no significant shift in the distribution of the lengths as recovery proceeds;
therefore, the first alternative must be correct.

Effect of Randomization During the Cell Cycle

It can next be asked (1) whether bacteria respond differently
because of chance events during or immediately preceding the time
of the measurements; that is, are they long responders at one time
but short responders later, or (2) whether an individual bacterium
is committed for an extended period to a certain type of response;
that is, are bacteria short responders throughout their life span?
To test the response of an individual bacterium over its cell division cycle, each was tethered to the surface of microscope cover-
slips by antibodies to individual flagella, as described by Silverman and Simon (1974) and developed further by Berg and Tedesco
(1975). The tethering method permits attractant to be delivered
and washed out at various intervals, minutes or even hours apart.
Counterclockwise rotation and clockwise rotation of a tethered cell
(observed along the axis from the flagellum to the body) are identified with tumbling and smooth swimming, respectively (Larsen et
al., 1974; Khan et al., 1978).

Table 2 shows the results of repetitive stimulation of five
tethered bacteria in the same microscopic field, with the attrac-
tant·α-methylaspartate. The attractant was delivered five times
and removed completely between stimulations. The delivery of large
amounts of attractant suppressed counterclockwise (tumbling mode)
rotation in each bacterium and caused a uniform clockwise (smooth
swimming mode) rotation for several minutes, with eventual recovery
to intermittent clockwise and counterclockwise rotation. All bacteria increased to nearly double their mass during the measurements,
yet the data show that individual sensitivities to the stimuli were
maintained. The same experiment was performed with six bacteria in

Table 1.   Independence of recovery times and position in the cell
           division cycle

Body length (μm)	Fraction of population of body length indicated with > 95% smooth swimming	Fraction of population of body length indicated with < 10% smooth swimming
<1.33	0.00	0.00
1.33–2.00	0.05	0.05
2.00–2.67	0.24	0.21
2.67–3.33	0.32	0.33
3.33–4.00	0.26	0.28
4.00–4.67	0.09	0.10
4.67–5.33	0.04	0.03
>5.33	0.00	0.00

The lengths of ST171 bacterial bodies were measured on projected films from the tumble frequency assay (Tsang et al., 1973). 186 individual bacterial lengths were measured from photographs at times for which less than 5% of the bacteria had recovered from the serine stimuli to define the distribution designated >95% smooth swimming. 212 individual bacterial lengths were measured after more than 90% had recovered to establish the distribution in the last column.

a nongrowth medium. These bacteria also maintained their tendencies to be either long or short responders to α-methylaspartate. Thus, whether growing or not, individual bacteria retain characteristically different sensitivities to stimuli for extended periods; the cells do not continuously randomize with respect to their chemotactic behavior.

The Cause of the Individualistic Variations

The above experiments establish that a bacterial culture that is genetically homogeneous and grown in homogeneous nutrient conditions nevertheless produces characteristically different individuals. These individuals retain their properties throughout their life cycle, showing that individuality is not a momentary effect of random bombardments from a fluctuating environment. Moreover, individuality is not a function of position in the cell division cycle.

How could such individuality be produced? A bacterium is a small cell. In the division and replication of cellular components certain molecules may well be present in such small amounts that

Table 2.   Response times (s) of individual bacteria on repetitive
           stimulation in growth conditions

Bacterium	A	B	C	D	E
Stimulus time (min)					
15	237	224	190	165	136
30	254	164	126	176	134
60	>300	265	191	136	145
75	>300	248	217	144	143
90	>300	253	172	152	148*
Average clockwise (smooth) response(s)	>278	231±18	179±15	155±7	141±3

Bacteria of strain SL3625, a leaky *fla AIII* mutant which
produces small numbers of flagella, were tethered with flagella
antibody to a coverslip by a modification of the procedure of
Silverman and Simon (1974). Bacteria were grown as described
(Tsang et al., 1973) except 200 μM ρ-hydroxybenzoic acid (ρHBA)
was added to the growth medium to improve motility (unpublished
results of J. Bar-tana, B. Howlett and D. E. Koshland). The
coverslip with attached bacteria was used to cover an open temporal
gradient observation cell of the type described by Macnab and
Koshland (1972). The observation cell was previously filled with
the wash medium to allow a liquid seal to form immediately the
coverslip was put in place (time 0 min). The behavior of bacteria
in one microscopic field at X 500 magnification was recorded on
videotape as follows. Cells were washed extensively (5 min at 12.7
chamber volumes per min) with the growth medium (without ρHBA); at
the times indicated the output line to the observation cell was
switched to a line containing this same medium +10 mM α-methylaspartate and a 15-s pulse of 12.7 chamber volumes of attractant was
delivered. The resulting immediate long clockwise rotation interval of an individual bacterium was measured as its "clockwise
response." After 5 min, the line was switched back to medium not
containing the attractant, and a 1-min pulse of 12.7 chamber
volumes per minute delivered. Each bacterium in the field which
was rotating for all stimuli was analysed by repetitive playback
of the videotape. The doubling time of SL3625 in this medium is
112 min. Mean responses are reported ± 1 s.e.m.

*Bacterium E divided 20 s after the delivery of the stimulus
at 90 min.

they are subject to Poissonian variation. In general, the standard
deviation (s.d.) of a Poisson distribution is the square root of n,
where n is the mean number of events. If there are 10^4 molecules
of particular type, the s.d. would be 100 molecules, giving a 1%
deviation. If, on the other hand, only 100 molecules are produced,
a 10% deviation would arise. In many cellular processes, <100
molecules participate (e.g., there are only 10-20 molecules of lac

repressor per cell (Gilbert and Müller-Hill, 1967), and enzyme
cofactor concentrations are extremely low (A. Wilson, 1962). In
other processes there may be more molecules overall but their ultimate number may have been determined by a small number of generating molecules (e.g., there are only 6-14 trp mRNA molecules per
bacterial cell, but each mRNA molecule is translated an average of
20 times (Baker and Yanofsky, 1970). Probabilistic fluctuations
in small numbers of molecules have been discussed as possible
causes of bacteriophage burst size variation (Delbrück, 1945) and
bacterial β-galactosidase concentration variation (Novick and
Weiner, 1957; Maloney and Rotman, 1973; Benzer, 1953).

To explain the role of random fluctuations and their relationship to cell responses, we must next consider the sensing apparatus
that controls bacterial behavior.

The Bacterial Sensing System

The bacterium has developed a sophisticated machinery involving receptors, a processing system, and a motor response, which is
analogous in many ways to the sensory systems of higher species.
Diverse models of this system and the data on which they are based
have been reviewed extensively elsewhere (Springer et al., 1979;
Adler, 1969; Koshland, 1977, 1981; Macnab, 1979; Hazelbaur and
Parkinson, 1977; Taylor and Lazlo, 1981); however, it seems appropriate to summarize and evaluate the information on the bacterial
model that is relevant to higher species.

Figure 3 is a schematic representation and description of a
general signaling system. Simply stated, it shows a series of
receptors that receive the primary signal and a processing system
that converts the primary signal into a concentration level appropriate for the response regulator, which in turn controls the
motor response. The roles of heredity, environment, and chance
are apparent in this system. The entire machinery is coded in the
genetic system of the bacterium; therefore, it might be argued
that the environmental responses are programmed in advance. For
example, the bacterium has approximately 30 receptors that respond
to individual chemicals, depending on the specificity of the receptor. If there is no receptor for a particular chemical, the bacterium is simply inert to that species. This response is exactly
parallel to those of our smell and taste organs, which are "blind"
to certain chemical stimuli because they lack receptors for them.
Moreover, the intensity of the response to an individual stimulus
depends on the affinity constants and the number of receptor molecules. Salmonella, for example, is far more sensitive to serine
than to ribose and more sensitive to ribose than galactose. In
the latter case, it has clearly been shown that the differing
sensitivities result from a larger number of ribose receptor molecules than galactose receptor molecules (Fahnestock and Koshland,
1979).

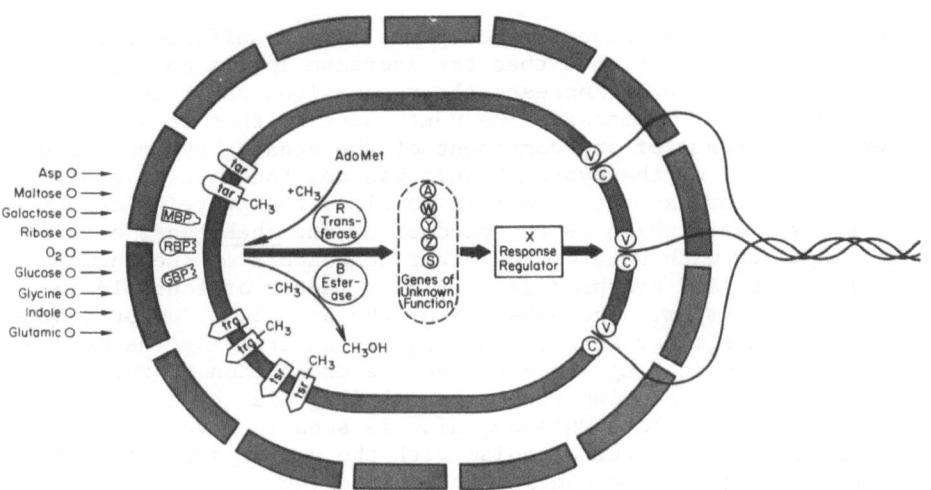

Fig. 3. Schematic summary of the bacterial sensing system. A variety of chemoeffectors (O_2, aspartate, glucose, etc.) pass through pores in the outer membrane and react with receptors in the periplasmic space (GBP = galactose-binding protein; RBP, ribose-binding protein; MBP, maltose-binding protein) and with membrane-bound receptors (tar for aspartate, tsr for serine). The membrane-bound molecules are methylated by the cheR gene product, which utilizes S-adenosylmethionine (AduMet), and demethylated by the cheB gene product. The peptides formed by the A, W, Z and Y genes are cytoplasmic and are known to be essential for the signal processing but have no specifically identified functions as yet. They, with R and B, control the level of the response regulator X.

Recently we have been able to modify the sensing system by introducing cloned genes to increase the copy numbers of individual components of the system (De Franco and Koshland, 1981; Wang and Koshland, 1980). One of the cloned genes has been the aspartate receptor. In using it we have enormously increased the sensitivity of the bacterium to aspartic acid. That finding supports the argument made above that the sensitivity to various chemicals can be established in a hierarchical order by producing more of a particular receptor and less of another. These studies showed further that the numbers of an individual receptor are not the only factor determining its potency. For example, the bacterium responds far more strongly when the receptor for aspartate is half-saturated than when the receptor for ribose is half-saturated, even though the aspartate receptor is present in only approximately one-fifth the amount of ribose receptor. The reason for this difference is that the aspartate receptor is more efficiently coupled into the remaining part of the processing system. Thus, we conclude that

the mere count of the number of receptors is not sufficient to
determine their potency, but that the increase in the numbers of a
particular receptor will increase the responsiveness of that organism to a particular chemical. The next question that we asked was
whether the increase of any component of the sensing system heightened the response of the system. This was not found to be the
case. It was, for example, possible to place the protein, carboxy-
methyl transferase, which is the product of the cheR gene, onto a
plasmid together with some of the other components of the chemotactic machinery and overproduce them by a factor of fourfold in
the biological system. The behavior of the organisms do not change
appreciably. A similar result was observed in studying the behavior of a mutant which depressed the entire production of the chemotactic machinery, including the bacterial flagella, by a factor of
approximately eightfold. This organism is seen to have essentially
identical responses to the organism with the overproduced machinery.
If either the absolute concentrations of the sensing system or the
relative concentrations of each of the components were crucial to
the proper functioning of the system, neither of these results
would have been obtained. It is clear, therefore, that certain
components of the system can be increased, either by themselves or
in conjunction with other components, without serious alteration
of the properties of the system. This might not be surprising in
a biochemical pathway in which some enzymes are apparently present
in excess. If that were so, increasing their concentrations so
they were more in excess would not change the rate-determining
step of that pathway. However, in a finely tuned sensory system,
alteration of any of the components might be expected to cause
more change than was actually observed.

Alterations Which Sensitize the System

The above statements should not lead one to conclude that all
modifications of the system are tolerated with little change in
the overall response. In fact, there are some very dramatic
effects. Mutations in a single gene in the sensing system can
alter the sensitivity to mixtures of chemicals. In most cases,
there was additivity or desensitization to a combination of signals
(Rubik and Koshland, 1978). However, in some cases there was an
enormous potentiation of one response by a second if a mutation in
one component of the sensing system occurred. This would mean
that a variation, one component of the sensing system, will make
the resultant bacterium enormously sensitive to environmental
changes. This result has now been extended to a number of other
mutants which in some cases completely lose the property of adaptation when exposed to an attractant after one of the components of
the sensing system has been modified. Such a result offers the
potentiality for enormous changes in the individual's responses.
The gene or its product could be modified either during heredity
or by post-translational processing. The resultant individual

then would have an extraordinarily different response to environmental stimuli, which would then be processed by other cellular
components to extenuate the differences.

These results were further followed up by deliberately introducing double mutations; i.e., a second mutation in another part
of the processing machinery after the first mutation had been
obtained. These changes allowed further insight into the system
(De Franco and Koshland, 1982; Stock et al., 1981). In some cases,
the second mutation apparently neutralized the effect of the first
mutation. In other cases, it accentuated it enormously. In some
of the cases, the second modification which neutralized the changes
caused by the first mutation did so for only a very limited range
of responses. The bacterium behaved apparently normally under
certain stress conditions or behaved very differently under other
conditions; in other words, the compensatory second mutation only
compensated for some of the functions of the bacterium, and allowed
the others to remain extraordinarily changed.

Environment, Heredity, and Chance

The sum of all these observations leads to a combined influence of environment, heredity, and chance on the responses of an
individual bacterium. Bacteria deliver offspring in most cases in
which genes are unchanged from generation to generation. The
hereditary components would appear, therefore, to lead to individuals which respond identically, other factors being similar. The
bacterium can be grown in an environment which is constant, which
is unchanging, from one bacterium to another. Yet, even under
circumstances of identical genes and identical environments, bacteria appear to be different, and this element is due to chance,
the Poissonian roll of the dice, which gives some bacteria more of
one particular part of the processing system and other bacteria
less. This individuality will be created even when environment
and heredity are maintained constant.

Of course, individuality can further be altered by hereditary
changes; mutations can occur in the bacteria, and these can lead
to dramatically different types of cellular responses.

Roles of Heredity, Environment, and Chance in the Responses of Cells

The number of receptors and hierarchic value system of an
individual cell is thus a composite of genetically determined constant responses, environmentally determined growth conditions, and
random fluctuations. The latter properties are also genetically
determined, since it is clear that the capacity for induced receptors is programmed in the genetic code. It may not be so obvious
that random fluctuations are programmed in the genetic apparatus,
but this is logical.

In many cases the receptors are made in the quantities of 10^4 copies per cell. The Poissonian variation from 10^4 gives a standard error of 1% of the total number of molecules, a variation which would be extremely minor among the total responses of the bacteria. However, if the total number of receptors is determined by two messenger molecules instead, the Poissonian variation in responses would be very great (2 ± 1.4). As mentioned in the first section, different bacteria have different responses, and these responses are characteristic of the bacterium over its entire life span. Thus, a genetic apparatus that dictates the number of molecules per cell at a key moment in development can build variation or determinism into individual responses. Some responses are subject to Poissonian variation because the number of messengers or final products per cell is very low, and other responses are deterministic because the number is very high.

We believe that this general response regulation model may very well hold for most responses of cells to stimuli at any level. Mammalian as well as bacterial cells may have systems in which there is a regulator for each response, the level of which varies around a threshold value to activate or inhibit a particular response. The level of the response regulator then depends on the composition of the cell as determined by heredity, by the growth conditions of the cell, and by chance factors at the time of division or throughout life.

The Possible Advantages of Programmed Randomization in the Responses of a Cell

If the above explanation for the individuality of a cell is correct, it is clear that any property could be programmed either to avoid or to have statistical fluctuations. If statistical fluctuations exist, it is presumed that they remain because (a) they have no deleterious effect on the organism, (b) they have a selective advantage, or (c) they cannot be eliminated without more serious adverse effects. It is therefore worthwhile to examine each possibility.

First, there seems to be many properties that can vary without serious effect on the cell. Once motility is assumed, the precise number of flagella seems relatively unimportant. Precise sizes of cells, numbers of enzyme molecules in nonrate-determining steps, and many other characteristics could well vary with little effect on the phenotype.

Second, a number of advantages of variation are apparent. One can certainly argue that nongenetic variability aids in the survival of a population subjected to widely varying conditions during its lifetime. If bacteria have a single or monolithic reaction to a particular chemical gradient, they might migrate as

a colony into a toxic situation or fail as a colony to be sufficiently sensitive to an essential new nutrient. If individual
bacterial variation occurs in a genetically homogeneous population, fractions of those bacteria might be either supersensitive
or insensitive. These bacteria would normally not survive as effectively as the bulk of the average, optimized population. They
would then have lower survival probabilities in most circumstances,
an unimportant loss if the main body of the colony survived. If,
on the other hand, the toxic and lethal situations described above
arose, a small fraction, on the wings of the probability distribution, might constitute the only bacteria to survive. These bacteria would then reproduce the species for future generations.
Genetic variation would not accomplish the same purpose since
selection in a rare toxic environment would produce a mutant poorly
adapted to more common conditions. However, during evolution populations are selected for survival in all of the widely varying
conditions of the environment. Thus, nongenetic variability would
be the preferred mechanism for accommodating random fluctuations in
the environment, and genetic variability the preferred mechanism
for accommodating to long-lasting environmental changes. The same
argument could be made for cells in a differentiated individual,
which must also be subject to selection both in growth and steady
state situations.

In a multicellular organism, random differences in cells might
be important in achieving differentiation. A gentle gradient, as
would exist in most cases during growth, would be unlikely to cause
dramatic changes between cells. But if otherwise identical cells
differed from each other by chance variation in certain response
regulators, the combination of dramatically different levels in
response regulator acted on by gradient could lead to a quantum
change in development. The differentiated cell could then excrete
special chemicals to alter its neighbors by potentiation or differentiation. The development of stalk and spore forming cells in
slime molds, of drones and workers in bee colonies, and of pecking
orders in higher species could easily be explained by such processes (E. O. Wilson, 1975).

Properties such as altruism, which might be selected against
in a genetic mechanism but which offers positive advantages for
survival of a colony, could be maintained by programmed randomness.
In that way, in times of prosperity colonies would have a range of
individuals working for the common good, with a few highly agres-
sive individuals who could be contained with little damage. In
times of adversity only the very aggressive might survive, but
they would breed a new generation with statistically random altru-
istic and aggressive individuals able to take advantage of prosperity. Other examples of properties of a cell population that
would be aided by random fluctuations have been discussed elsewhere
(Silverman and Simon, 1974).

An example of an adverse effect that may be compensated by other advantages could be the total number of proteins elaborated by a cell. If there is an optimal level, the production of some proteins might have to be limited. In that case, some function in which random fluctuation caused suboptimal performance might be more than compensated by a vital function without fluctuation. The organism would produce excess molecules for the vital function and leave the number of molecules for the less vital function acceptable but marginal.

APPLICATION TO HIGHER SPECIES

Although higher species operate by an interaction of many cells, it has been found that effects within certain cells can be dominant, particularly in sensing hormonal functions. For example, it is believed that schizophrenia may be due to an excess of dopamine or of dopamine receptors. Parkinsonism, on the other hand, appears to be due to a deficiency of dopamine and dopamine receptors. Thus, in these cases, despite correct "wiring" of different receptor cells to each other, the system operates abnormally because some of the individual components are excreting excess amounts of a neurotransmitter, or insufficient amounts of a neurotransmitter. Similarly, it has been shown that deficiencies of hormones in certain stages of the life cycle can cause abnormal behavior and insufficient responses.

It is being found in more and more cases that certain patterns of behavior, which have in the near past been considered to be entirely environmental, can now be produced by hereditary deficiencies. Schizophrenia, manic depressive, alcoholism and aggressiveness are each identified with a genetic lesion in careful studies using Swedish adoption groups (De Fries and Plomin, 1978). Moreover, in many cases, the syndrome can be cured or alleviated by treatment with chemicals, a suggestion that it is a neuronal defect, and not a particular wiring diagram problem which causes behavioral abnormality.

If we now combine what we observe for the individual cell with that for the whole organism, we can at least rationalize a number of the different individual responses which are so important in human behavior. In the first place, except for identical twins, no two human beings appear to have identical genetic components. Thus, individuals begin with the potential for different processing systems which control behavior. This does not automatically establish individuality, since it is at least conceivable that differences in the level of certain constituents will have no influence on behavior (the analogy to the cases in which we can introduce additional cloned genes without alteration in behavior). However the likelihood that at least some responses would be different is very high. Even if the individuals have identical genes, the statistical fluctuations could result in different behaviors. And, in fact, concordance in behavior between identical twins is not absolute. In schizophrenia, for example, if one twin becomes schizophrenic, the second twin has a 50% chance of being schizophrenic also (Kety et al., 1975), a high correlation but not 100%.

Another potentiality for individuality arises from the alterations in the genetic components in the sensing system. If the response regulator model holds for mammalian neurons as it seems to do, it is conceivable that one parent might have a slightly deficient amount of one component of the system which will not affect his behavior; in the other parent, a slightly deficient amount of a second component of the system which will not appreciably affect her behavior. The combination of the two components, however, will produce an individual of serious aberrant behavior. If this behavior, for example, produced the kind of potentiation which we mentioned in the mutant, it will mean this individual will be highly susceptible to certain environmental signals, particularly when they were delivered in the presence of other signals. It would thus potentiate responses which in some cases might be highly desirable, such as heightened senses of smell or taste, while in other cases they might be highly undesirable, such as tendencies to respond violently to small stimuli.

There is a further element of chance introduced by the interactions of a sensing system with outside stimuli at different points in the life cycle. At various times, such as puberty or menopause, when the hormonal balances are changing, certain individuals may have deficiencies of hormone levels so that external stimuli have abnormal responses, responses which would not be observed at other phases in the life cycle. The combination of a chance component of environmental stress at a critical period of the life cycle in an individual with an unfortunate conformation of genes is a further component of individuality.

CONCLUSION

The study of the molecular aspects of heredity in a very simple organism suggests that all three components of heredity, environment, and the chance Poissonian distribution of molecules in a cell can affect the individual. The alteration of these components offers enormous numbers of permutations, so that each individual will respond differently to the same stimuli in a certain number of cases. Understanding of these permutations and their influence on behavior can provide important insight to the variation in an individual's responses to his environment.

REFERENCES

Adler, J., 1969, Chemoreceptors in bacteria. Science 166, 1588–1597.

Aswad, D., and Koshland, D. E., Jr., 1974, Role of methionine in bacterial chemotaxis. J. Bacteriol. 118, 640–645.

Baker, R., and Yanofsky, C., 1970, Transcription initiation frequency for the tryptophan operon of Escherichia coli. Cold Spring Harbor Symp. Quant. Biol. 35, 467–470.

Benzer, S., 1953, Induced synthesis of enzymes in bacteria analyzed at the cellular level. Biochim. Biophys. Acta 11, 383–395.

Berg, H. C., and Brown, D. A., 1972, Chemotaxis in Escherichia coli analyzed by three dimensional tracking. Nature 239, 500–504.

Berg, H. C., and Tedesco, P. M., 1975, Transient response to chemotactic stimuli in Escherichia coli. Proc. Natl. Acad. Sci. USA 72, 3235–3239.

Cairns, J., 1960, The initiation of vaccinia infection. Virology 11, 603–623.

Delbrück, M., 1945, The burst size distribution in the growth of bacterial viruses (bacteriophages). J. Bacteriol. 50, 131–135.

DeFranco, A. L., and Koshland, D. E., Jr., 1981, Molecular cloning of chemotaxis genes and overproduction of gene products in the bacterial sensing system. J. Bacteriol. 147, 390–400.

DeFranco, A. L., and Koshland, D. E., Jr., 1982, Construction and behavior of strains with double mutations in chemotaxis genes. J. Bacteriol. 150, 1297–1301.

DeFries, J. C., and Plomin, R., 1978, Behavioral genetics. Ann. Rev. Phys. 29, 473–515.

Engelmann, T. W., 1881, Neue Methode zur Untersuchung der Sauerstouffausscheidung pflanzlicher und thierischer Organism. Pflueger Arch. Ges. Physiol. 25, 285–292.

Fahnestock, M., and Koshland, D. E., Jr., 1979, Control of the receptor for galactose taxis in Salmonella typhimurium. J. Bacteriol. 137, 758–763.

Gilbert, W., and Möller-Hill, B., 1967, The lac Operator is DNA. Proc. Natl. Acad. Sci. USA 58, 2415–2421.

Hazelbauer, G. L., and Parkinson, J. S., 1977, in Receptors and Recognition, Chapman & Wall, Random, New York.

Iino, T., et al., 1975, Temporary expression of flagellar phase-1 in phase-2 clones of diphasic Salmonella. J. Gen. Microbiol. 89, 265–276.

Kety, S. S., Rosenthal, D., Wender, P. H., Schulsinger, R., and Jacobsen, B., 1975, in Genetic Research in Psychiatry, Johns Hopkins Press, Baltimore.

Kahn, S., Macnab, R. M., DeFranco, A. L., and Koshland, D. E., Jr., 1978, Inversion of a behavioral response in bacterial chemotaxis: explanation at the molecular level. Proc. Natl. Acad. Sci. USA 75, 4150–5154.

Koshland, D. E., Jr., 1977, A response regulator model in a simple sensory system. Science 196, 1055-1063

Koshland, D. E., Jr., 1981, Biochemistry of sensing and adaptation in a simple bacterial system. Ann. Rev. Biochem. 50, 765-782.

Larsen, S. H., Reader, R. W., Kort, E. N., Tso, W. W., and Adler, J., 1974, Change in direction of flagellar rotation is the basis of chemotactic response. Nature 249, 74-77.

Lederberg, J., and Iino, T., 1956, Linear inheritance in transduction clones. Genetics 41, 743-757.

Macnab, R. M., 1979, Bacterial motility and chemotaxis: the molecular biology of a behavioral system. Crit. Rev. Biochem. 5, 291-341.

Macnab, R. M., and Koshland, D. E., Jr., 1972, The gradient-sensing mechanism in bacterial chemotaxis. Proc. Natl. Acad. Sci. USA 69, 2509-2512.

Maloney, P. C., and Rotman, B. J., 1973, Distribution of suboptimally induced β-D-galactosidase in Escherichia coli. J. Mol. Biol. 73, 77-91.

Novick, A., and Weiner, M., 1957, Enzyme induction as an all-or-none phenomenon. Proc. Natl. Acad. Sci. USA 43, 553-566.

Pfeffer, W., 1883, Locomotorische richtungsbewegungen durch chemische Reize. Ber. Dtsch Bot. Gest. 1, 524-533.

Powell, E. O., 1958, An outline of the pattern of bacteria generation times. J. Gen. Microbiol. 18, 382-417.

Quadling, C., and Stocker, B., 1957, The occurrence of rare motile bacteria in some non-motile Salmonella strains. J. Gen. Microbiol. 17, 424-436.

Quadling, C., and Stocker, B., 1962, Environmentally-induced transition from the flagellated to non-flagellated state in Salmonella typhimurium: the fate of parental flagella at cell division. J. Gen. Microbiol. 28, 257-270.

Rubik, B. A., and Koshland, D. E., Jr., 1978, Potentiation, desensitization, and inversion of response in bacterial sensing of chemical stimuli. Proc. Natl. Acad. Sci. USA 75, 2820-2824.

Schaechter, M., Williamson, J. P., Hood, J. R., Jr., and Koch, A. L., 1962, Growth, cell and nuclear divisions in some bacteria. J. Gen. Microbiol. 29, 421-434.

Silverman, M. R., and Simon, M. I., 1974, Flagellar rotation and mechanisms of bacterial motility. Nature 249, 73-74.

Spiegelman, S., 1951, The particulate transmission of enzyme-forming capacity in yeast. Cold Spring Harbor Symp. Quant. Biol. 16, 87-98.

Springer, W. R., and Koshland, D. E., Jr., 1977, Identification of a protein methyltransferase as the cheR gene product in the bacterial sensing system. Proc. Natl. Acad. Sci. USA 74, 533-537.

Springer, M. S., Goy, M. F., and Adler, J., 1979, Protein methylation in behavioral control mechanisms and in signal transduction. Nature 280, 279-284.

Spudich, J. L., and Koshland, D. E., Jr., 1975, Quantitation of the
    sensory response in bacterial chemotaxis. Proc. Natl. Acad.
    Sci. USA 72, 710–713.
Spudich, J. K., and Koshland, D. E., Jr., 1976, Non-genetic indi-
    viduality: chance in the single cell. Nature 262, 467–471
Stock, J. B., Maderis, A. M., and Koshland, D. E., Jr., 1981, Bac-
    terial chemotaxis in the absence of receptor carboxylmethyla-
    tion. Cell 27, 37–44.
Stocker, B. A. D., 1949, Measurements of rate of mutation of flagel-
    lar antigenic phase in Salmonella typhimurium. J. Hyg.
    (Camb.) 47, 398–413.
Taylor, B. L., and Lazlo, D. J., 1981, Perception of Behavioral
    Chemicals, Elsevier, New York, in press.
Tsang, N., Macnab, R., and Koshland, D. E., Jr., 1973, Common
    mechanism for repellents and attractants in bacterial chemo-
    taxis. Science 181, 60–63.
Tso, W. W., and Adler, J., 1974, Decision making in bacteria: a
    chemotactic response of Escherichia coli to conflicting
    stimuli. Science 184, 1292–1294.
Wang, E. A., and Koshland, D. E., Jr., 1980, Receptor structure in
    the bacterial sensing system. Proc. Natl. Acad. Sci. USA
    77, 7157–7161.
Wilson, A., 1962, Regulation of flavin synthesis by Escherichia
    coli. J. Gen. Microbiol. 28, 283–303.
Wilson, E. O., 1975, Sociobiology, Harvard University Press,
    Cambridge.

DISCUSSION

RUTTER: I was interested in the problem of arguments derived from very small single cell organisms like E. coli to a multicellular organism like the human being, not so much with respect to the extension of the Poissonian principle but the extension, for example, of the notion of regulation by small numbers of molecules. It is true in an organism having the volume of the bacterium that when one talks about pH one is considering very few proton molecules within the cell, thus your argument is cogent. On the other hand, the mammalian cell is hundreds of times bigger than a bacterial cell. So, under those circumstances that particular entity is present in hundreds of times the concentration and the argument of Poissonianism based on a few molecules does not necessarily hold.

          One can extend that and say, well, there are some regulatory molecules that are present in a few molecules per cell. But if that were the case, then perhaps you'd like to see in higher organisms differences in the affinities with which such small molecules bind their so-called receptors. And, I guess maybe that's true, but even in cases where the highest affinities are known, 10^{-9} to 10^{-14} perhaps or that order of magnitude, one still gets hundreds of molecules inside mammalian cells. And so, I think that, pushed to the limit, you would postulate on this simple model that there are some unknown molecules which exist in relatively small concentrations. And I wonder whether there isn't another postulate which is rather an easier one to develop, and that is that there is a Poissonian distribution of cells. It's not that there's simply a few molecules, but a few cells.

KOSHLAND: What makes them different then?

RUTTER: The numbers of cells between individuals. Let's say, for example, that you have ten cells, especially in the brain, which regulate a secondary process for which there are many cells-- hundreds of thousands of cells. Then the aggregate output of an entity in a multicellular system is really the units in that entity. Or, you could make elaborate modifications of that. That still allows the Poissonian idea to be extended but doesn't depend upon a small number of molecules.

KOSHLAND: I think there are two answers to that. Let's just take the small number of cells. If they are not different within themselves, then they have to be different due to their environment. That is, an initial cell is formed; it then duplicates and the central cell has five cells around it which cover it. Now, the outside cells have a different diffusion problem than the center cell. Now, that is really what we thought about in terms of bacteria. So we can talk about variation due to the different environmental gradients that identical cells are subjected to.

The second argument relates to the cells. If there is not a change in the environment there can be an exponential distribution in the development of cells. This has been shown by Smith and Martin and many others. The best way to get Poissonian distribution is by statistical fluctuations of a small number of molecules. You can get Poissonian variations by a very complex structure which generates fluctuation and get really what physicists call chaos but that requires an enormously clever mixture of a large number of enzymes and velocity. And it seems to me that the smart way to do that is just by having a few molecules. You can easily have a small number of having a thousand molecules in a mammalian cell reacting with another thousand molecules as a combination. It is interesting that statistical fluctuation occurs on going through the cell cycle.

LOWENSTEIN: There is another way one might get fluctuations-- and this may be a mechanism more and more important as one gets to the larger cells of eukaryotes: topological variations of the reactant molecules. After all, what matters isn't necessarily the overall concentration of the reactants in the aqueous phase of the cell but often--and certainly so in membrane-bound reaction systems-- the important thing is how much of the molecule is located at a particular site. And so, if one or more of the molecules is sitting on a two-dimensional matrix--a membrane--much fluctuation may arise from local geographical differences in the number of molecules. One lesson of cellular chemistry of recent years is how surprisingly few molecules are free to move inside cells. Even some small inorganic ions have a little difficulty. Take the calcium ion, for example. Whereas in water it is very mobile, in the cytoplasm it hardly is. Because of rapid and high-capacity calcium sequestering mechanisms everywhere inside the cell, this ion's domain of free diffusion is very restricted. The domains over which calcium ions can transmit a message inside the cells are not more than one micron or so in diameter (B. Rose & W. R. Loewenstein, Science 190:1204, 1975). Thus, where this ion is a critical reactant, it is its concentration in a small locale that matters. This is, of course, also true for macromolecules, particularly when they are membrane-bound. So, variations of topological nature may be a factor, perhaps the biologically more important one in eukaryotes.

KOSHLAND: Absolutely. I think that I would completely agree with you except as to which is more important. I am not sure we know. I would argue that the mechanism that Bill Rutter mentioned and the geometry you mention are mechanisms that are used. That is, you can make a bunch of identical cells and as they grow they are in different environments that will turn various enzymes on and off. What I am saying about this, is that there is an added factor that we must think about in terms of individuality. And, it could be tested in the following experimental way. There are people like Pardee who postulate there is a single protein acting like a repressor molecule which controls replication. If you could clone it and overproduce it, you ought to be able to synchronize mammalian cells. The converse of that is to say if you put mammalian cells in swirling cultures you will never get individuality because the individuality depends on a different environment. If you always did get distribution, it would mean that there's a Poissonian variation within the cell. I am saying that there is some biological advantage in that and that there are ways of testing the hypotheses.

LOEWENSTEIN: At one higher level of structure--if I may call it structure--yes, I would call it organismic structure, most of the fluctuations arise then through interacting cellular elements. Let us now interconnect two or three of your "individual" bacteria, to get the paradigm a bit closer to a higher organism, then we get interactive fluctuations and, perhaps, more important here for individuality, hierarchically interactive fluctuations.

KOSHLAND: Really, what we are doing in the bacteria is giving the building blocks for more complex systems. It's quite clear as you go to higher organisms you are going then to be dealing with hierarchical things which are much more complex. Let's just take a neuron and a bacterium. Neurons are in some ways essentially like bacteria; that is, they are getting information from other neurons by chemical signals (neurotransmitters) on receptors. And neuronal output, instead of being reversible flagellar rotation, is release of a neurotransmitter. And they process information with a neuron essentially the same way the bacterium does. They may not do so identically but very similarly. Neurons can integrate information from what is called inhibitory and excitatory inputs, which we would call repellants and attractants in bacteria and get an output which is an integrated sum of the input information. The difference of your brain and that of a lower species is not because of the way in which the different neurons work, it's that there are so many more of them. So individuality in the bacteria has all these elements and you can put the elements in various combinations and hierarchies in a complex species and give many more possibilities of variation.

STANFORD: As a nonscientist, I feel like a babe in the woods or a fish out of water, but not as a shrinking violet. I found this presentation extremely satisfying because I feel my own ego swimming toward this attractive gradient. In other words, you are confirming an intuitive bias which may not be intuitive at all; it may be programmed!

I want to ask you about one sentence that bothers me because it doesn't seem to be supported altogether.

You say that the roles of heredity, environment, and chance are apparent in this system. The entire machinery is coded in the genetic system of the bacteria. Therefore, it might be argued that the responses to the environment are programmed in advance. And yet, from your conclusion presented here and in your paper, the determinists might say that you have waffled, pardon the slang, by attributing an influential role to environment and Poissonian rolls of the dice, as you called it. I believe you attempted to explain developments, but I am still not persuaded that you needed to argue in behalf of environment and Poissonian dice rolls.

KOSHLAND: I'll tell you exactly what I meant. If you go through the major mental diseases, more and more people believe they have an innate biochemical basis. The evidence from Swedish adoption studies that some manic-depressive illness, suicides, schizophrenia, alcoholism are inherited, is really now very strong. These may be what I call single neuronal defects. That is, schizophrenia can be traced to excess of dopamine excretions, Parkinsonism, to too little dopamine, and so forth. The fact that we can cure some of these things with drugs, like manic-depressives with lithium, tend to add to this argument. In these adoption studies with Swedish children, the incidence of schizophrenia in the general population is one to two percent, whereas the incidence if you have one close relative with schizophrenia goes up to 10 to 12 percent. So it isn't 100 percent; it's 10 to 12 percent, which is what you'd expect from a multigenic situation. The environment does act as a factor. I think anorexia nervosa, in my opinion, is an innate imbalance problem. When I say innate, I mean something could happen to cause a damage in pregnancy or a chance combination of genes that I talked about before occurs. So, now you have an individual with a self-starvation syndrome. The psychiatrists have, in my opinion, two self-defeating propositions. One is "the child starves to death because he is trying to fight his parents, and it's the only way he can do it." The second argument is, "he loves his parents so much that he doesn't want to grow up," and "starvation prevents puberty." The facts are that one child has to be force-fed because he or she is going to starve himself to death and yet there are two other siblings who show no damage at all. It seems unlikely that one child would react so strongly to environment whereas two siblings would show no damage at all. It is much easier to believe there is something wrong with the brain and, after all, lack of appetite is seen in many illnesses. The child may actually be eating much more than he wants in order to please its parents, but he's still starving. Thus, I believe it is not all genetic but there is genetic predisposition. If you react abnormally to your parents telling you to go to school or to dress well, I call that in your genes. A lot of times mental illness is caused because you put stress on individuals at the time when their hormone balances are changing. That's chance,

so that's what I mean by a combination of chance and heredity. The
same stress on an individual who doesn't have a marginal genetic
situation would not cause any trouble. Some schizophrenics were
destined to be schizophrenic from birth. Some probably can be
saved with a protective environment. Some individuals will never
become abnormal no matter how terrible the environment.

TOBACH: I take a very different position in regard to the
analysis and interpretation of your data. I really am very pleased
that you are here because I use your work in my class and in my
lectures all the time, coming to a very different interpretation
than you do--so I think that's very interesting. And, I don't
really want to go into that now, but I do have some questions that
I always had that I want to ask you now. I think the work is very
elegant and very important.

          In the analysis of the flagellar response to chang-
ing internal and external conditions, internal conditions being
genetic or another kind of biochemical relationship, I think that
you have worked out, if I remember correctly, the actual biochemical, biophysical changes that the gene is participating in to
affect the movement of the flagellum. So that, in other words, in
your work you can give us a picture of the actual process whereby
that gene functions in regard to that flagellum. Right?

KOSHLAND: Yes.

TOBACH: Now, also, isn't it so that you could work out in a
test tube the energy changes that are related to the function of
that molecular configuration that you call a gene and the molecular
configuration that you call the flagellum in terms of what it does
and what it doesn't do?

          So, to repeat Dr. Loewenstein's up-with-reductionism
idea, we could reduce it down to that level without any difficulty.
Right?

KOSHLAND: Right.

TOBACH: Okay, so that in other words it is conceivable that
in your analysis of the response of the flagellar organization to
the so-called attractants and repellants, you could really define
all of those systems in regards to some kind of energy change, or
biochemical, biophysically described process.

          Okay, given that fact that we all have then and we
know now that what we're really talking about at this point of our
knowledge, the ultimate level is some kind of energy exchange that
goes on that brings about a new condition in regard to the movement
in molecules. I'd like to have Sid Fox read the piece that Dr.

Metz brought in by the president of Cornell University. Because
what I would like to remind us all of is that in everything we're
talking about, we're talking about changing the relationships of
systems that have evolved. What you might call co-evolved. I
think it is very important what the man says.

FOX: I'll be glad to read it again.

            The emphasis is that evolution is neutral. Refer-
ring to the theory of evolution, Rhodes (Johns Hopkins Magazine 35
5, 1982) says: "It is a theory of the mechanism of descent with
modification which seeks to explain how new species develop. It
does indeed contradict certain ideas of the manner of creation.
But like all scientific theories, it provides no ultimate interpretation of the origin of natural laws themselves, for it no more
proves them to be the result of random chance than it proves them
to be the servant and expression of purpose."

TOBACH: Now, that last sentence is the one that I am trying
to bring to our attention at this point. Because the evolution of
the relationship of matter to itself, which is what we're talking
about, we're really talking about it in terms of living inanimate
systems, the changes that take place have a certain fitness to
them. And this has something to do with the way change takes place
in general. To impute to those changes and to those co-evolving
systems some kind of purposiveness is a very big step to take and
I think that to go even further from the molecular level, that you
are talking about two very complex societal processes like mental
disease is again another big step in the other direction. So that
if we can understand your work in terms of the appropriateness of
co-evolving systems so that the systems that we're looking at are
still here for us to look at and didn't disappear someplace, I
think we have another way of looking at your data without getting
involved in determinism and extrapolation to other systems before
we know how the other systems work.

FOX: The way it looks to me is that the early evolving
system did not disappear. It evolved partly by climbing hierarchical stairways. Essentially, what I am saying is that molecular
determinism, the fit of molecules (a subject for which Koshland is
the pioneer), led to the genetic system and to genetic determinism.
(T. H. Morgan, E. O. Wilson), which in turn fed into behavioral
determinism, as we shall be discussing and testing here. I think
molecular biologists know a lot about how genetic systems work and
I think molecular evolutionists have recently learned something
about how coded genetics originated on a narrow and unique evolutionary pathway (see, e.g., Follmann in Naturwissenschaften 69,
75-81, 1982).

KOSHLAND: Can I take a crack at answering Dr. Tobach? Because we obviously think alike in some ways but very differently in others. Let's just take the energy thing. I mentioned that S-adenosyl methionine is used to methylate and then is constantly demethylated. You may say that's crazy because the bacteria are swimming around ceaselessly looking for a new environment, but we calculated they spend only one to two percent of their energy just swimming around. The one thing that unites all biological systems, right up to man, is that we waste a tremendous amount of energy to get control. Think of your brain. Your brain uses 20 to 30 percent of your resting energy. You may say, well that makes a lot of sense; it's coordinating a lot of things. When you go to sleep, the energy requirement goes down very little. You don't use much more energy in order to think. A hibernating organism actually depresses its total metabolism. But all you have to do is be awakened once by a saber-toothed tiger or a fire in a life-threatening situation to recognize the value of not hibernating your brain. If you can respond immediately, you're alive. If you respond slowly like a hibernating bear, you're dead. Now, if it's easy to get enough food to operate your brain all the time so it can turn on and turn off rapidly, obviously the survival of the species is helped. And it looks to me as if all species spend a lot of energy to maintain control.

As far as evolution is concerned, my opinion comes back to this programmed behavior versus free will. The bacterium is programmed ahead of time to constantly search for a better environment. Birds are born with immediate desire to hunt for food. They don't have to be trained by their mother to do this. The bacterium doesn't know precisely what gradient they are going to experience and the bird doesn't know where it's going to find the worm, but both are programmed to react instinctively to predictable situations. So you have in every species enough programs to survive and then individual variation occurs in terms of learning its individual environment.

TOBACH: When you talk about something being programmed in advance, should I take that literally?

KOSHLAND: I mean that the genes are there so the bacterial chemotaxis system responds.

TOBACH: To a known situation?

FOX: To even an unknown situation.

KOSHLAND: Let me give you an example. A bacterium has a spectrum of receptors which does not allow it to detect every compound, but it's enough. For example, you're repelled by rotting food. The odors are not necessarily the compounds that are toxic, but

there is an association so that those odors that are generally
emitted by rotting food are repellent to you. So all you have to
do is get a signal of one repellent even though many of the compounds may be toxic.

TOBACH: No. But I'm asking you about the advance business.
In other words, you are saying that the genes are programmed in
advance--in advance of future events. There is something someplace
that prepares this organism with something for the future and the
thing that's preparing them is based on what kind of pre-knowledge?
I mean, how does that take place?

KOSHLAND: The DNA of the chromosomes . . . .

TOBACH: In the DNA. In the DNA the molecules there have
some prior information about future events.

KOSHLAND: Yes.

TOBACH: Now, that's very interesting. Can you give me some
material process whereby the DNA molecules know what's going to
happen? I mean, how does that happen?

KOSHLAND: Dr. Rutter will, I am sure, discuss the the expression of genes later, but . . . .

TOBACH: Oh, I know how genes are expressed, but I am asking
you a very different question. So it selects from something that
already exists.

KOSHLAND: The bacterium has a repertoire of enzymes designed
for its survival as in chemotaxis. If one of those is bad, those
species disappear and they disappear very rapidly. Now in the
laboratory I can keep a methionine auxotroph alive by feeding it
methionine. But in the wild it would die. So it is programmed to
meet future needs. Human beings are kept alive by getting vitamins. We don't make certain vitamins so we must eat them. But if
we get them, the organism will survive. Thus, we are programmed
in advance for survival.

TOBACH: All of that is absolutely true, but you see what I'm
asking is a slightly different question. Let me ask it again. In
other words, we have existing systems that have experienced certain
relationships in the past. Out of those systems, some have remained
and some have been lost for any particular organisms. Right?
Therefore, I do not understand how it is that these systems which
are the end product, the derivative of a whole series of things
that happened in the past are now going to tell us something that's
going to happen in the future.

KOSHLAND: I think you have raised an interesting question
about the biology of anticipation.

TOBACH: Right. Preadaptation is one of the big arguments
that goes on in evolution theory.

KOSHLAND: What kind of systems are developed in anticipation
of something in the future? Well, ostensibly, the biology of
anticipation is really based on history; that is, it's probabilistic that an organism will meet a certain set of challenges, characteristic of the environment.

TOBACH: Based on the past.

KOSHLAND: Not only based on past specific circumstances per
se, but on the necessity to respond to some class of changes which
has existed in the past, and, therefore, likely to be a required
response in the future.

TOBACH: Yes. But you see what you're saying is that what is
the past of these organisms is now going to be more than just a
range of past experiences because, in effect, what you are saying
is that they have within them a whole series of responses and
receptors, whether molecular or whatever, and now they are going
to be doing something into the future. Okay. What I'm saying is
that system does not make any provision for new species and new
changes to things that never existed before.

KOSHLAND: Oh yes, they do. The answer is the argument that
there's a standard rate of mutation in all species.

TOBACH: Mutation means change.

KOSHLAND: Mutation means change. Right. And that is built
into the systems because you can anticipate future change and,
therefore, you build in the selection. I'll give you an example.
Years ago, gonorrhea organisms which did not resist penicillin and
various other drugs were essentially nonexistent in the population.
They now exist in large numbers. Over a period of time with the
massive use of penicillin, selection occurs for organisms that
resist penicillin. Changes in the environment, changes in the ice
age, and other changes select for mutation.

TOBACH: What you told me and all that Dr. Rutter has said,
is that things change.

KOSHLAND: Sure.

TOBACH: That's great. When you use the word mutation,
that's all you're saying. And all that I am saying is that it is

not necessary to impute some predetermined program for these things
to happen if you accept the fact that everything changes.

MONROY: I use the immune system to bring in the story of
another organism because you have been discussing bacteria. Well,
the organism I want to bring in is much more obligate than bacteria
--that is, the trypanosome.

         This is a story which is just emerging recently,
and which may answer all the questions raised by Dr. Tobach. If I
may take five minutes.

         The story is with the number of trypanosomes in the
blood of the infected animal or man for that's the agent of the
sleeping sickness and increases to a peak and then all of a sudden
drops to a new level, and this because of the immune reaction
developed by the host organism, which kills most of the infecting
organisms.

         Now, let's look at the problem not from our side but
from the side of the infecting trypanosome. Should the trypanosome
have the ability to completely overcome the defense, pretty soon it
would kill the organism and then it would find no organism on which
to feed.

         On the other hand, if the defense of the organism
were so strong as to kill the trypanosome, the outcome would be
the same. After a few days of relapse, you get a new wave of trypanosomes in the blood, and that goes on periodically for a number
of days. That was known since the beginning of the century.

         Then comes the fascinating story. Each change, each
peak, each fluctuation is due to the fact that a new variant of the
organism arises which is not sensitive to the immune response. The
story is that the antigenic property of this creature is due to a
glycoprotein coat that surrounds the animal and gives the immunological identity to this trypanosome. Each variant has a different
composition.

         Now, the story goes that it has been discovered that
when they analyze the amino acid sequence of the coat protein, each
strain is different. So each variant protein is coded by a different gene. Under condition of environmental stress, one gene is
silenced and another gene takes over. And it has been calculated
that every organism has an endorsement of about a hundred controlling genes. What is more interesting is if you start from an
original stock, a variant.

The evidence that the variants originate under the environmental stress is demonstrated by the fact that if you injected a clonal culture of trypanosomes into an animal which has been immunosuppressed, then you don't get any variants. So it is the strength of environment.

What is interesting is that, if you start to follow variant A, you don't get just a random variant set. You get a certain specific variant, which is B. You get C, not 100 percent but with a high percentage, and then C,D. At one point you take the variant D and you inject back into the tsetse fly; you go back to A.

So now you have a series of genes which probably have originated by duplication of an ancestral gene and which conditions the response of the organism to the environmental stress. When a variant arises, you have a rearrangement of the genome very similar to the rearrangements of the genome you get in yeasts for the mating types.

This is one of the more fascinating stories of how the environment can impinge on a genetic system.

THE EVOLUTION OF INDIVIDUALITY AT THE

MOLECULAR AND PROTOCELLULAR LEVELS

Klaus Dose

Institute for Biochemistry
Johannes Gutenberg University
Mainz, Germany

INTRODUCTION

The most important bioelements (= organoelements) hydrogen, carbon, oxygen and nitrogen, are also the most abundant elements throughout the Universe besides helium, neon, and silicon (Fig. 1). In the Universe carbon is about four times as abundant as silicon. Certainly, the abundance of elements in various celestial bodies may vary greatly depending on the history of these celestial bodies.

The cosmic abundance of the four most common bioelements, like that of all other elements, is directly related to the physical properties of the neutrons, protons, and other particles of which the nuclei of these elements have once been formed. Since the beginning of the Universe, matter has been subjected to environmental strains which have caused it to adapt by evolutionary processes. The extreme heat and pressure in the interior of stars, e.g., have caused the interconversion of matter by nuclearchemical reactions. The physical laws of these processes always have controlled the abundance of elements in the Universe. In particular the giant stars steadily emit significant portions of their matter into space, sometimes in a very conspicuous way, as in the case of novae and supernovae phenomena. The interstellar clouds formed by these emissions of stellar matter may contract again to yield new stellar systems. Also, our Solar System has evolved from such an interstellar cloud (presolar nebula) about five billion years ago.

All chemical elements (with the exception of the noble gases) readily interact chemically whenever the conditions are favorable. Spontaneous formation and interconversion of molecules, therefore, occur in a variety of cosmological settings. The processes are

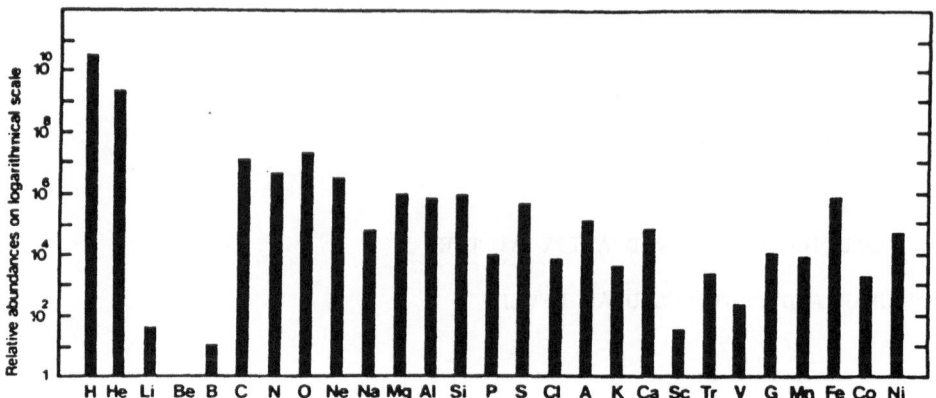

Fig. 1. Interstellar abundances of elements. The abundances of elements in the Universe is largely of the same order of magnitude. Source: A. G. W. Cameron (1970), Space Science Reviews 15, 121-146.

usually referred to as chemical evolution. Even interstellar clouds contain a wealth of spontaneously formed molecules. Most of the chemical compounds so far observed and identified are carbon compounds (Table 1).

Analogous hydrogen compounds of silica have not been detected. The diversity of carbon compounds, their stability and cosmic abundance suggest that living systems, if they exist elsewhere in the Universe, are likely based on the chemistry of carbon compounds as shown for all terrestrial organisms. Also, water is likely the standard solvent for other (extraterrestrial) forms of life. The dependence of living systems on water as a solvent finally roots in the cosmic abundance of water, its chemical stability, and its physical-chemical fitness. Ammonia could theoretically compete with water as a biological solvent. But its thermal and photochemical lability as well as the narrow temperature range, within which it exists in the liquid state (-78°C to -33°C at 1 atm pressure), make it less suitable.

After water (about 80%), the proteins are the dominant constituents (about 15%) of the interior of almost all living cells. Therefore, the question now arises whether the crucial functional and structural role of proteins, their abundance and biological fitness, are also rooted in a chemical or even geochemical and cosmochemical matrix. The answer that we are able to give today strongly supports this notion.

Table 1. Observed interstellar molecules in dense molecular clouds

INORGANIC		ORGANIC
H_2 hydrogen OH hydroxyl SiO silicon monoxide SiS silicon sulfide NS nitrogen sulfide SO sulfur monoxide	DIATOMIC	CH methylidyne CH^+ methylidyne ion CN cyanogen CO carbon monoxide CS carbon monosulfide
H_2O water H_2H^+ H_2S hydrogen sulfide SO_2 sulfur dioxide	TRIATOMIC	CCH ethynal HCN hydrogen cyanide HNC hydrogen isocya- nide HCO^+ formyl ion HCO formyl OCS carbonyl sulfide
NH_3 ammonia	4-ATOMIC	H_2CO formaldehyde HNCO isocyanic acid H_2CS thioformaldehyde
	5-ATOMIC	H_2CHN methanimine H_2NCN cyanamide HCOOH formic acid HC_3N cyanoacetylene
	6-ATOMIC	CH_3OH methanol CH_3CN cyanomethane $HCONH_2$ formamide
	7-ATOMIC	CH_3NH_2 methylamine CH_3C_2H methylacetylene $HCOCH_3$ acetaldehyde H_2CCHCN vinyl cyanide HC_5N cyanodiacetylene
	8-ATOMIC	$HCOOCH_3$ methyl formate
	9-ATOMIC	$(CH_3)_2O$ dimethyl ether C_2H_5OH ethanol HC_7N cyanotriacetylene

Amino acids, the building blocks of proteins, consist of the elements H, C, N, and O, whose cosmical abundance has already been emphasized. Amino acids have been extracted from various carbonaceous meteorites in amounts of several 10 μg per g (Table 2). These amounts are far above the values expected on the basis of equilibrium thermodyanmics. Amino acids have even been extracted, though in much smaller amounts (ng/g), from lunar fines. Moreover, amino acids have been extracted from the oldest known sedimentary rocks (Isua sediments, about 3.8 billion years old) and even from most recent vulcanite materials. Amino acids are also the predominant products in laboratory experiments simulating the chemical evolution of organic compounds from the constituents of the atmosphere and the hydrosphere of the primitive Earth. These constituents likely included H_2O, CO_2, CO, N_2, and smaller amounts of NH_3, CH_4, and H_2, but no free O_2. Whenever such mixtures (gaseous or liquid) are exposed to energy pulses such as electric discharges, ultraviolet light, ionizing radiation, heat or shock waves, amino acids will be found among the reaction products (Table 3). The first experiment of this kind has been performed by S. Miller in the fifties. This experiment was an excellent pioneer experiment that stimulated myriads of consecutive experiments, although it was designed (from our present point of view) in a naive way: the atmosphere of the primitive Earth was never identical with the $CH_4/NH_3/H_2O/H_2$-atmosphere used by Miller and his closed glass apparatus shows no geochemical relevance.

As yet, we cannot satisfactorily explain why amino acids are mostly the predominant reaction products and why they exhibit such a pronounced cosmochemical, geochemical and biological fitness. We are sure, of course, that their fitness is closely related to their zwitterion character, their high melting (decomposition) points, their low vapor pressure and their generally good solubility and stability in O_2-free water.

Table 2. Amino acid compositions of extracts from carbonaceous chondrites (non-biological amino acids excluded)

Amino Acid	Murchison	Murray	Nagoya
	(per cent of total amino acids)		
Aspartic acid	3.4	5.5	10.1
Glutamic acid	6.6	4.8	20.3
Glycine	33.6	17.7	27.6
Alanine	14.0	6.6	7.8
α-Aminoisobutyric acid	19.4	50.7	0
β-Alanine	6	5.7	11.9

Table 3. Relative amounts of some amino acids and related compounds produced from various atmospheres with electric discharges, X-rays, β-rays, and heat

| Products[a] | Molar Ratio (Glycine = 1) | | | |
	Electric Discharges (sparking) (CH_4, NH_3, H_2O, H_2)	X-rays (CH_4, CO_2, H_2O, NH_3, N_2, H_2)	β-rays (CH_4, NH_3, H_2O, H_2)	Heat (CH_4, NH_3, H_2O)
Glycine	1.00	1.00	1.00	1.00
α-Alanine	0.54	0.25	2.0	0.17-0.83
β-Alanine	0.24	0.20		0.52
α-Aminobutyric acid	0.08			0.17
Aspartic acid	<0.01	0.50		0.13-0.63
Urea	<0.01	>100	>100	
Acetate	0.24	1.25		
Lactate	0.49	0.75		

[a] A wider variety of compounds was actually obtained in all experiments.
Source: Fox and Dose (1977).

There are certainly numerous other biologically significant compounds that have been produced in experiments simulating chemical evolution. These compounds include various sugars, lipids, heterocycles such as purines and pyrimidines. But nucleotides, the building blocks of our nucleic acids have never been produced in a properly designed simulation experiment. This lack of evidence makes it presently very difficult to directly relate the individuality of living organisms to that of their evolutionary precursors. The information for the individuality of living organisms is to a significant degree stored in their genetic apparatus, but it is expressed by converting (transcribing) the genetic information into the information of proteins (Fig. 2). However, we can present evidence suggesting that the individuality of the evolutionary precursors of living cells (we shall call them protocells or sometimes precells) is largely rooted in the individual polypeptides (protoproteins) abiotically formed from prebiotic amino acids. The consequences of this thesis regarding protobiological evolution shall also be discussed at the end of this paper. The various evolutionary stages and sequences referred to so far are summarized in the following Fig. 3.

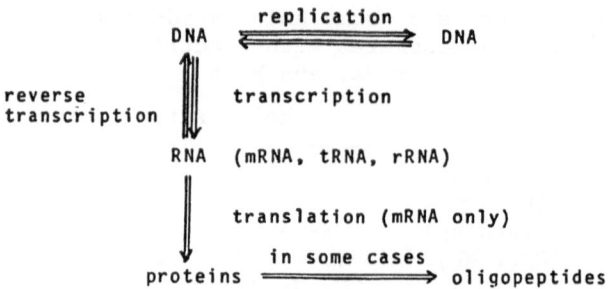

Fig. 2.   Flow of biological information.   Replication = reduplica-
tion of DNA.   Transcription = transcribing the nucleotide sequence
of a DNA into the corresponding sequence of an RNA.   Translation =
transfer of the sequence of nucleotide triplets (trinucleotide
sequences) into the amino acid sequence of a protein or polypeptide.

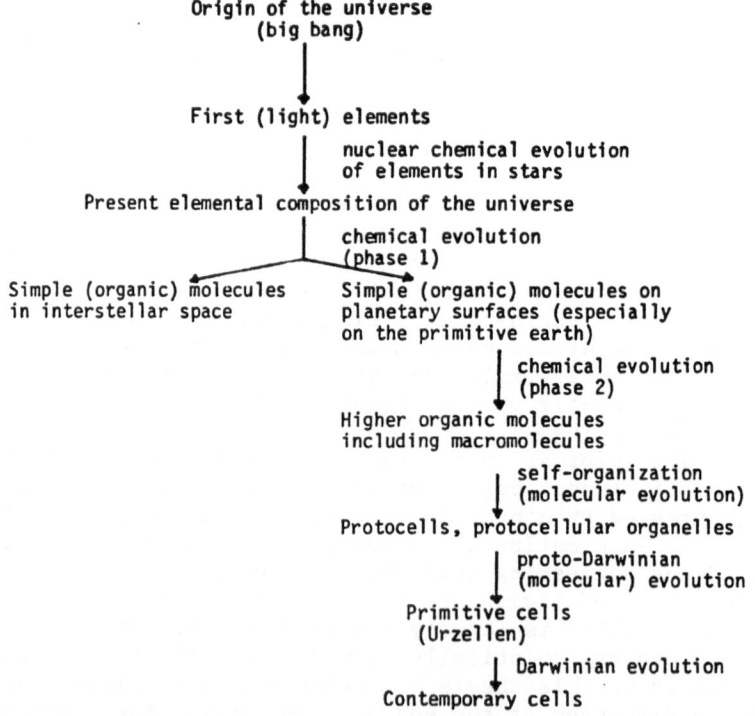

Fig. 3.   Evolutionary sequences from the origin of the Universe
to contemporary cells.

The evolutionary theory of the origin of life suggests that the gap between chemical and Darwinian evolution was bridged by proto-Darwinian evolution (or molecular evolution). Many experimental results, largely obtained by laboratory investigation, allow us to draw a detailed picture of the various stages during chemical evolution. It is largely agreed that these processes finally yielded protocellular systems on the borderline between nonlife and life. The subsequent (proto-Darwinian or molecular) evolution of proto-cellular systems to the phylogenetic precursors of our contemporary living systems, however, is a controversial issue. Like chemical evolution also, proto-Darwinian evolution is not an object of direct analysis. The possible steps can only be reconstructed in the laboratory by simulation experiments. As yet, no fossilized remains of proto-life in old sediments have been identified beyond doubt, whereas many more recent stages of Darwinian evolution can be well identified with the help of the fossil record and the comparative analysis of contemporary species.

Much of the present controversy on the origin of contemporary life concerns the origin of genetic information. What came first: the proteins or the nucleic acids? The two opposing views may be compared with the ancient question: What was first, the hen or the egg? Contemporary biology or biochemistry cannot directly answer these questions. The view that life has begun with a spontaneously formed self-replicating nucleic acid looks intriguing. But this view is, at least at the present, without chemical foundation. The disadvantage of the proteins-first theory, however, is seen in the view that a flow of information from proteins to nucleic acids is "forbidden" by the central dogma of molecular biology (Fig. 2); indeed, as yet, no experimental evidence allows us to conclude that the ribosomal protein synthesis, the translation process, can be reversed. But this process has evolved from more primitive precursors which likely were operative on a reversible basis. Numerous observations in molecular biology confirm the view that proteins can recognize specific nucleotide sequences or can recognize built-up new nucleotide sequences in the absence of a coding nucleic acid matrix. Nonrandom polymers of amino acids (protoproteins), in contrast to polynucleotides, are relatively easily formed in prebiotic simulation experiments. In the following, it will be demonstrated why and how these amino acid polymers gain their individuality and why they are nonrandom and informational.

CONDENSATION OF AMINO ACIDS TO OLIGOMERS AND POLYMERS (PROTOPROTEINS)

In dilute aqueous solution, the chances for a thermal condensation of biomonomers (amino acids, monosaccharides, nucleotides) to polymers (peptides or proteins, polysaccharides, or nucleic acids) are rather small for thermodynamic reasons (substantial positive free energy change).

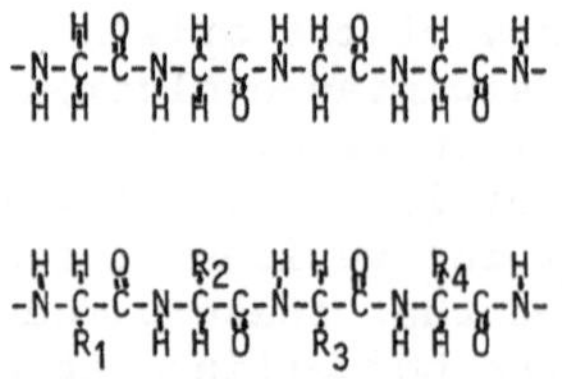

$$-N-C-N-C-C-N-C-C-N-C-C-N-$$

$$-N-C-C-N-C-C-N-C-C-N-C-C-N-$$

Fig. 4. Proteins are polypeptides. Any protein is formally a
polysubstituted polyglycine. Above: polyglycine. Below: partial sequence of a polypeptide (protein).

For instance, in order to synthesize a polypeptide from amino
acids a condensation reaction between the reactant amino acids
under elimination of water has to take place. In the laboratory
the chemist usually "activates" the amino acids by converting them
into more energy-rich derivatives. In the living cell this activation is achieved by a primary reaction of amino acids with adeno-
sine triphosphate. In any event, the synthesis of a peptide
requires the removal of water and the addition of free energy
according to the following equation:

$$H_3N-\underset{R_1}{\overset{+H}{C}}-COO^- + H_3N-\underset{R_2}{\overset{+H}{C}}-COO^- \quad \xrightarrow[+ H_2O]{- H_2O} \quad H\,N-\underset{R_1O}{\overset{+H}{C}}-C-N-\underset{H}{\overset{H\,R_2}{C}}-COO^-$$

$\Delta G^{o'} = 8-20 \text{ kJ/mole}$

The amount of free energy, $\Delta G^{o'}$, in the formation of higher
peptides is somewhat lower because the energy required to increase
the distance between the H_3N^+-group (at the N-terminal end of the
peptide) and the $-COO^-$-group (at the C-terminal end) is lowered
with the increasing length of the peptide chain. But the overall
equilibrium remains so unfavorable that solutions 1M in each amino
acid would yield at equilibrium a 10^{-99} M concentration of protein (M.W. 12,000). The thermodynamic barrier is surmountable by
removal of the water formed as a byproduct when peptide bonds are
synthesized. This can be achieved by evaporation of the water (at
elevated temperatures) or by binding the water chemically with the
help of condensing agents; in the prebiotic realm minerals, cyan-
amide and related compounds could have acted as condensing agents.

Areas with temperatures of 65°-180°C probably were more ubiquitous four billion years ago than they are now. Laboratory experiments have shown that condensation of amino acids at elevated
temperatures is by far more efficient than the processes involving
chemical binding of the water released during the condensation.

More recent experiments have even shown that polymers of amino acids can be produced with substantial yields in an aqueous system at moderate temperature and pH when water-soluble carbodiimdes are used as condensing agents.

Laboratory experiments, in particular those by S. W. Fox and his associates (Fox et al., 1959; Fox and Dose, 1977), have shown that mixtures of amino acids can be polymerized to a variety of individual, nonrandom macromolecules (proteinoids). The problem of nonrandomness in polypeptides or proteins shall be explained in the following paragraphs.

Polypeptides (or proteins) are formally polysubstituted poly-glycines (Fig. 4).

The sequence of the amino acids (characterized by the residues R_1, R_2, R_3, R_4, etc.) in the peptide chain of a protein determines its individual structure and function (e.g., as enzyme, hormone, or transport protein). It can be said that the amino acid sequence of a polypeptide or--more general--the nonrandom sequence of the various monomers in biopolymers (proteins, nucleic acids) represents a certain amount of information or individuality. The terms information or individuality are used here with respect to molecular interactions. To inform or to express individuality means in molecular biology the transfer of a message from system A to system B and to cause system B to react in a certain way. All molecules contain a distinct amount of information or individuality (expressible by the sum of their chemical and physical properties).

Proteinoids (protoproteins or abiotically formed polypeptides) may have a distinct informatory content if their building blocks, the amino acids, and their reaction products, are ordered in a non-random fashion within the polymer. The origin of this order is found to a large degree in the chemical properties of the reactant amino acids. This fact shall be elucidated by the following play-ing-card model.

We choose three different playing cards instead of three dif-ferent amino acids (with the side chains R_1, R_2, and R_3), we may combine them to sets of three playing cards in $3^3 = 27$ different ways (Fig. 5). Also, three different amino acids may theoretically be combined to yield 27 different tripeptides. However, the vari-ous amino acids have, in contrast to playing cards, a different spatial structure and chemical reactivity. The result is that there exist favorable and unfavorable combinations. Fig. 6 shows in a simplified way the formation of a favorable combination (tripeptide).

It can indeed be shown by experimentation (Nakashima et al., 1977; Hartmann et al., 1981, 1982) that the sequence of building

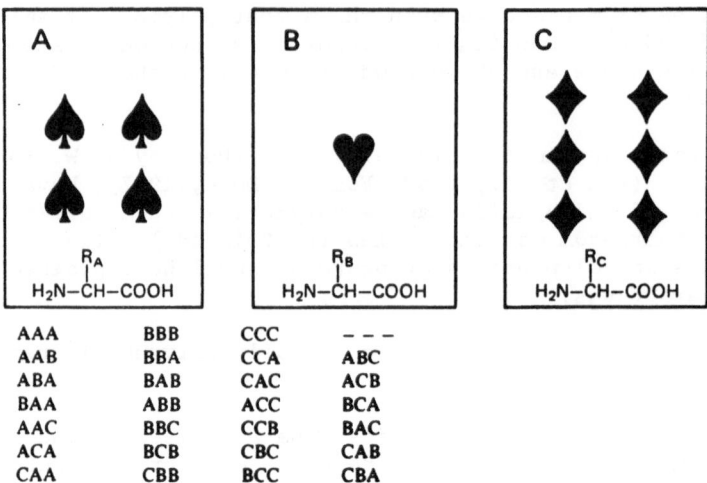

Fig. 5. Random association of playing cards when forming "triplets" (all of the same shape and affinity).

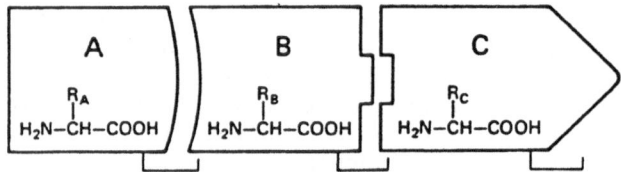

Fig. 6. Nonrandom association of amino acids when forming a tri-peptide (each amino acid of unique shape and reactivity). In actuality, nonrandomness is not total. The uniqueness of shape of amino acids is exaggerated, but is more accurate than using cards to indicate amino acids.

blocks in abiotically formed polypeptides (proteinoids) is con-trolled by the chemical properties of the reacting amino acids and that their reactivity is in turn also influenced by their environ-ment (pH, temperature, concentration, presence of minerals and other reactive components including the growing polymer).

The model suggested in Fig. 6 is, therefore, backed by experi-mental evidence.

It is long known that amino acids also partially decompose at elevated temperatures. But even this process is favorable with

respect to the formation of catalytically active proteinoids.
Heinz et al. (1979) have shown that amino acids during the thermal
process also yield pterines, flavins and other heterocyclic compounds which are simultaneously incorporated into the polymer. It
has also been shown that in the presence of glutamic acid the formation of certain tripeptide sequences with pyroglutamic acid as
N-terminal amino acid is favored (Nakashima et al., 1977; Dose et
al., 1981). Fig. 7 shows a section of a suggested structure of a
proteinoid; various chromophores are crosslinked with short peptide chains of definite amino acid sequences. Such polymers would
correspond to the chromoproteins (or chromoproteides) of contemporary cells. If they are discussed as evolutionary precursors of
contemporary proteins these proteinoids could be called protopro-
teides.

More evidence is required to corroborate the inference that
proteinoids were evolutionary precursors of contemporary proteins
and genes. But many proteinoids synthesized in the laboratory
have at least some of the properties which we would have to ascribe
to the evolutionary precursors of proteins. These are:

1. catalytic (enzyme-like) activities
2. the property to form individual cell-like structures
   (precells, microspheres) by self-organization (self-
   assembly)

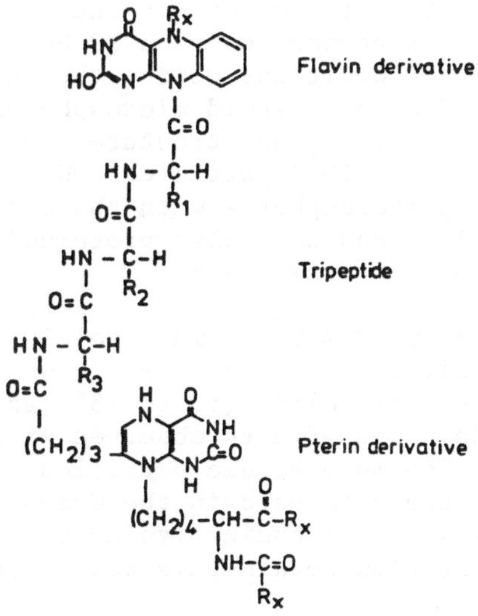

Fig. 7. Postulated partial structure of a thermal copolyamino
acid (proteinoid, protoprotein).

Some catalytic properties of various proteinoids are summarized in the following table (Table 4). In context with the formation of flavins (redox-systems) and pterines, which is of particular significance in lysine-rich amino acid mixtures, redox reactions such as the reductive amination of α-ketoglutaric acid and the oxidative deamination of glutamic acid deserve special attention.

The second property of proteinoids, their ability to form individual superstructures (precells or microspheres) by self-organization, shall be discussed in the next section.

SELF-ASSEMBLY OF PROTEINOIDS INTO CELL-LIKE STRUCTURES

For the first time, Fox et al. (1959) have described the formation of "spherules" during cooling of saturated (aqueous) proteinoid solutions. Later, the spheric, cell-like particles have been called (proteinoid) microspheres (Fig. 8). Proteinoids rich in acidic amino acids (aspartic acid, glutamic acid) are generally good starting materials for microsphere production. The process is surprisingly simple and quick. To some degree proteinoid microspheres have structures similar to the "organized elements" in carbonaceous chondrites (meteorites) or ancient sediments. But at least the structures in meteorites have very likely a non-aqueous origin. The size of the proteinoid microspheres depends on the composition of the solution and the nature of the polymer. They show osmotic properties; that is, they swell and shrink in dependence on the salt concentration. During these processes their diameter may vary by a factor of 3. The average diameter of proteinoid microspheres is between 0.5 and 80 μm, depending on the starting material and the conditions during the self-assembly process. A "population" of proteinoid microspheres is relatively homogenous with respect to size and structure. One gram of proteinoid may yield as much as 10^{10} particles. When comparing the properties of proteinoid microspheres with those of bacteria, Fox [in Fox and Dose (1977)] found that also proteinoid microspheres can be differentiated by the Gram stain.

Tests indicate that microspheres made solely from acidic proteinoids are Gram-negative. Gram-positive microspheres, however, are formed from a mixture of acidic (50 to 75%) and lysine-rich (25 to 50%) proteinoids. The effects observed with this stain are not simply explained. The more acidic particles do not bind crystal violet, one of the dyes used in the Gram stain, and which is itself basic. Only when the basic (lysine-rich) proteinoid is incorporated do the particles retain this basic dye.

The envelopes of these microspheres are in a crude sense semipermeable inasmuch as larger organic substrates such as glycogen are more readily kept than smaller molecules, e.g., glucose.

Table 4. Summary of important catalytic activities of thermal
polymers of amino acids

Substrate	Authors	Active amino acids in polymer	Activity [μM/mg/min]
Hydrolysis			
p-Nitrophenyl acetate	Noguchi & Saito (1962); Rohlfing & Fox (1967); Usdin et al. (1967)	His, Lys, Asp-imide, Glu, Asp Lys, His, Ser, Tyr	1.5×10^{-3}
p-Nitrophenylphosphate	Oshima (1968)	Basic and neutral amino acids, Asp-imide	2×10^{-3}
ATP	Fox & Joseph (1965)	Zn, lysine-rich polymer	$\sim 10^{-4}$
Decarboxylation			
Glucose or glucuronic acid	Fox & Krampitz (1964); Hardebeck & Fox (1967)	Lys	$\sim 10^{-4}$
Pyruvic acid	Krampitz & Hardebeck (1966)	Glu, Thr	$\sim 10^{-4}$
Oxaloacetic acid	Rohlfing (1967)	Lys	0.3
Amination and Deamination			
α-Ketoglutaric acid H_2N-donor: urea and others	Krampitz et al. (1967)	Lys and other basic amino acids	2×10^{-3}
Other α-ketoacids H_2N-donor: glutamic acid	Krampitz et al. (1968)		$\sim 10^{-4}$
Oxidoreductions			
H_2O_2 (catalase reaction)	Dose & Zaki (1971a,b)	hemoproteinoid Phe inhibitor	1
H_2O_2 and H_2-donor (NADH, guaiacol) (peroxidase reaction)	Dose & Zaki (1971a,b)	Lysine-rich hemoproteinoid most active	0.2

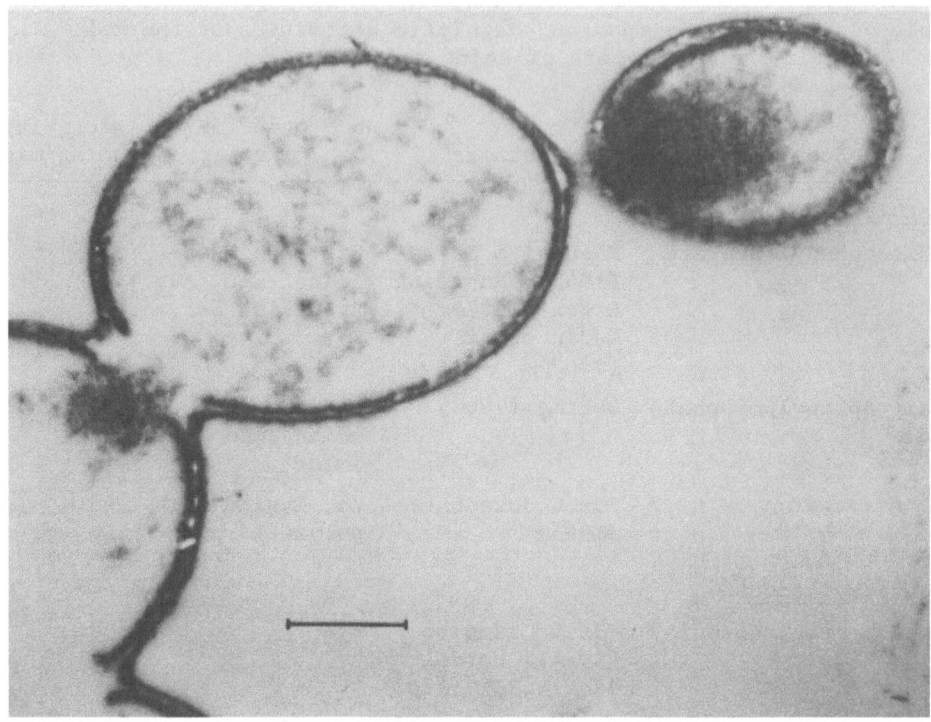

Fig. 8. Electron micrograph of an osmium tetroxide-stained proteinoid microsphere after pH has been elevated. Double layer in boundary is prominent. Source: S. W. Fox, Miami, Florida.

     Proteinoids retain their individual catalytical properties in
microspheres. Thus metabolic reactions occur in proteinoid microspheres if suitable substrates are offered. Accordingly, proteinoid microspheres meet some criteria for a living system. Although
microspheres are mainly composed of individual macromolecules, their
individuality is not only a superposition of the properties of their
building blocks as many observations testify.

     Microspheres may propagate by budding. The buds appear spon-
taneously on the surface of microspheres allowed to stand in their
mother liquor. The buds can be released from their "parent" microspheres and may grow to full-sized microspheres by aggregation.
Because of these cell-like properties proteinoid microspheres have
been proposed as models for protocells [Fox and Dose (1977)]. At
the present we cannot decide whether proteinoid microspheres are
the correct models for precells or even protocells. An answer
would appear within reach if it could be shown that proteinoid

microspheres could evolve to cells which store their genetic information in nucleic acids.

REMAINING QUESTIONS

The experimentation with proteinoids has led to at least three important questions:

1. Do the proteinoids synthesized in the laboratory (and interpreted as models for prebiotic proteins or protoproteins) contain the "correct" information; that is, the information required for the evolution of prebiotic systems to the first living cells?

2. If the information stored in proteinoids is a kind of protogenetic information (an information that preceded the contemporary genetic information which is stored in nucleic acids), how could that information become translated into the sequence of nucleotides in nucleic acids, especially DNA?

3. How did the first nucleic acids originate?

Some molecular biologists and theoreticians have postulated that the genetic information of living systems has ab initio evolved in polynucleotides. The problem is that polynucleotides are not self-replicating. They need enzymes for "self-replication." The chances for a prebiotic formation of polynucleotides and specific polynucleotide replicases are extremely small.

Two important reasons which have been brought forward to oppose the genes-first view (see also the introductory section) are summarized below:

1. As already stated, there is no self-replication of nucleic acids in contemporary cells. A catalyst is required. In contemporary cells the catalyst is a specific nucleic acid replicase. Like the synthesis of all other enzymes also the synthesis of this enzyme is coded by DNA and RNA.

,2. Even if there exists an (as yet only suggested) non-enzymic, but sufficiently specific, catalysis of polynucleotide replication, where did the first replicable polynucleotide come from? To be replicable, the polynucleotide must have the same very specific structure as contemporary polynucleotides (D-2-deoxyribose or D-ribose in 3- and 5-position esterified with phosphoric acid and in 1-position β-glycosidically linked to a specific N-atom of a purine or pyrimidine base; see also Fig. 9).

An efficient and largely error-free replication of polynucleotides is only possible if the polynucleotide has exactly the

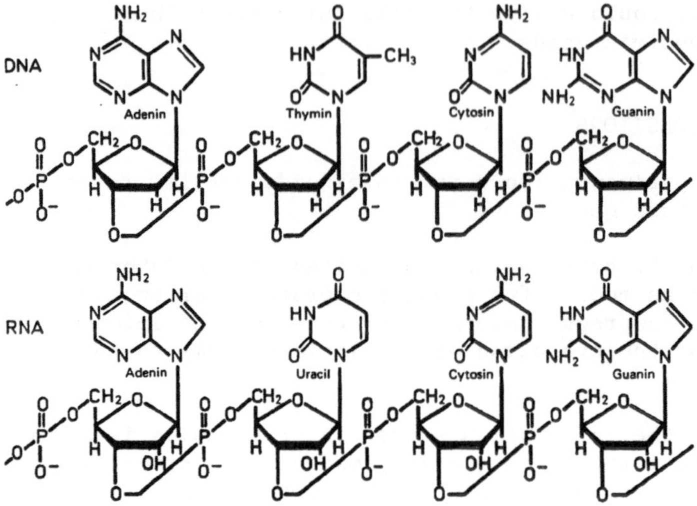

Fig. 9. Nucleotide sequences (of model fragments) of DNA and RNA.
All linkages are strictly stereospecific.

stereochemical structure presented in Fig. 9. Only then the proposed evolutionary sequence shown in Fig. 10 (right) would be a
realistic theory. However, because of these specific stereochemical requirements (prebiotic selection and conservation of chiral
structures!), the spontaneous (abiotic or prebiotic) formation of
perhaps decanucleotides appears extremely unlikely at the present.
The same applies to the spontaneous formation of other biopolymers
with highly ordered structures such as polysaccharides (cellulose,
glycogen, starch, and others).

In summary, the question how individuality evolved at the
cellular or protocellular level including the question of polyamino acids-first (protoproteins-first) or polynucleotides-first
is still under discussion and experimentation. Prebiotic simulation experiments back the protoproteins-first alternative (left in
Fig. 10), but the widely held view that polynucleotides came first
(right in Fig. 10) can more conveniently be related to contemporary
molecular biology although it is as yet not backed by prebiotic
simulation experiments.

Both alternatives are presented in Fig. 10 within the framework of relevant evolutionary sequences.

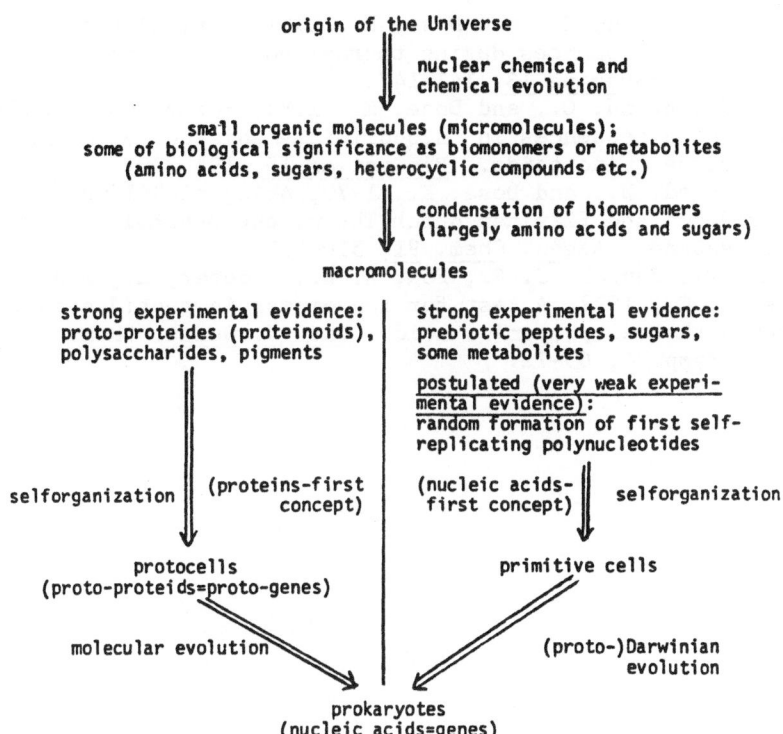

Fig. 10. Scheme of evolution of prokaryotes: According to the
"proteins-first concept" (left) proteinoids (proto-proteides =
proto-genes) were the first informative biopolymers. During further evolution, their information became translated by a "reverse
translation" into the nucleotide sequences of first nucleic acids
(now forbidden by the "central dogma of molecular biology").
According to the "nucleic acids-first concept" (right) ab initio
biological information has evolved from spontaneously formed polynucleotides (proto-genes). This concept is as yet not backed by
simulation experiments on prebiotic evolution.

REFERENCES

Fox, S. W., and Dose, K., 1977, Molecular Evolution and the Origin
    of Life, rev. ed. Marcel Dekker, New York.
Fox, S. W., Harada, K., and Kendrick, J., 1959, Production of
    spherules from synthetic proteinoid and hot water. Science
    129, 1221-1222.
Fox, S. W., Harada, K., and Vegotsky, A., 1959, a thermal polymer-
    ization of amino acids and a theory of biochemical origins.
    Experientia 15, 81-84.

Hartmann, J., Brand, C., and Dose, K., 1981, Formation of specific
    amino acid sequences during thermal polymerization of amino
    acids. BioSystems 13, 141-147.
Hartmann, J., Brand, C., and Dose, K., 1982, Formation of specific
    amino acid sequences in carbodiimide-mediated polymerization
    of aqueous amino acids. BioSystems 15, 195-200.
Heinz, B., Ried, W., and Dose, K., 1979, Abiogene Bildung von
    Pteridinen and Flavinen durch Thermische Behandlung von
    aminosauren. Angew. Chem. 91, 510-511.
Nakashima, T., Jungck, J. R., Fox, S. W., Lederer, E., and
    Das, B. C., 1977, A test for randomness in peptides isolated
    from a thermal polyamino acid. Int. J. Quant. Chem. Quant.
    Biol. Symp. 4, 65-72.

DISCUSSION

TOBACH: Do you think that the two theories of proteins-first
and of polynucleotides-first are really mutually exclusive?

DOSE: We have no experimental evidence for the spontaneous
formation of polynucleotides on the primitive Earth. There is the
possibility that the first polynucleotide was formed by a proteinoid. Also, some modern proteins exhibit self-directed polymerase
activity. You may have heard about the activity of Qβ replicase.
Sumper and Luce (Proc. Nat. Acad. Sci. 72, 162-166, 1975) have
studied it in Eigen's laboratory. They have presented evidence
indicating that Qβ replicase synthesizes RNA which has, at least
in some parts, nucleotide sequences that are identical with those
of the Qβ phage RNA. This kind of polynucleotide synthesis
occurs in the absence of the RNA-DNA matrix. It is principally
possible to use the information contained in the structure of a
protein (or protoprotein) for the synthesis of individual polynucleotides.

TOBACH: In your figure you showed there were some polypeptides also before you get the polynucleotides. So, those are the
ones that I was particularly focusing on as possibilities.

FOX: I think a good deal of the difficulty that usually
attends these questions is a kind of smog that is generated by the
loose use of the terms self-replication of DNA or self-replication
of RNA. There's not real evidence for such self-replication, if
one speaks strictly. Nucleic acids are replicated, and they are
replicated and transcribed and translated by hosts of enzymes. So,
for the nucleic acids to come into existence, at least in the
modern situation, the polyamino acids have to be there first. A
primary importance of the polyamino acids was first shown clearly
in the models for the primordial events. This emphasis in the
modern situation has been made most forcefully, rather recently,
by Arthur Kornberg. In a book titled DNA Replication (1980),
Kornberg points out that in the original Watson-Crick formulation
of replication there was no mention of enzymes, as follows:

> It was suggested in 1953 that A-, T-, C-, and G-contain-
> ing precursors might orient themselves as base pairs
> with a DNA template and then be "zippered" together
> without any enzyme action. However, to the biochemist
> it is implicit that all biosynthetic and degradative
> events are catalyzed by enzymes, making possible
> refinements of control and specificity, and rapid rates
> of reaction.

The work of Kornberg and others shows that the replication of DNA in E. coli depends upon numerous specific enzymes. We think the situation is now being desmogged slowly.

KOSHLAND: I notice the composition that you have for interstellar matter doesn't agree with the estimates of the atmosphere of the prebiotic Earth. Is Mars any closer in composition based on the spectroscopic evidence?

DOSE: Chemical evolution takes place both within open space in the Universe and on planetary bodies. Small planets, including Mars, have a secondary atmosphere that has been built up by volcanic outgassing from the interior. Spectroscopic data on the Martian atmosphere only indicate the presence of nitrogen, carbon dioxide, little oxygen and almost no water, but no organic molecules.

METZ: When it comes to basic individuality, amino acids are like the animals going into the Ark; there are two of each kind: D and L. Living organisms are composed only of L amino acids. Is there any simple explanation of why you can't have a life system of D amino acids forming peptides?

DOSE: No. The energy content of D and L amino acid is almost equal. So are their rates of formation in prebiotic experiments. It is generally believed that chirality is a typical sign of life. That's not the case. Matter appears to be basically chiral. I have asked many nuclear physicists about this problem. As yet, nothing is known about asymmetries within nuclear structures of chemical elements. But some related facts are known. For instance, during decay of strontium-90 the electron which is ejected spins primarily in one direction. For each spin around a given axis, there is a noncongruent mirror image of this spin; that is, any such spin is chiral. The β-decay of strontium-90 thus indicates asymmetric effects within its nucleus. There are other examples of prebiotic chirality: our galaxy has the shape of a spiral and any spiral is chiral.

METZ: If you heat D amino acids, do you get proteinoids?

DOSE: Yes, we also get proteinoids. The procedure to heat amino acids at temperatures about 100° in order to produce proteinoids has at least a disadvantage regarding racemization. Whenever one starts with pure L or D amino acids the products will be more or less racemized. Racemization is also a serious problem in living systems, in particular for long-lived proteins. Some of our body proteins are synthesized very early in our development and then they are just deposited; for instance, the dentin of our teeth and most proteins of our eye lenses. Here racemization takes place at a rate of around 0.1% per year. It's a relatively low rate, but after decades a substantial amount of amino acids will be D. If we are sixty or seventy years of age, around ten percent of these amino acids will then be in the D form.

HARDIN: An ineffective biological form?

DOSE: We don't know of any consequences. It has been specu-lated that this racemization has something to do with the aging of our lenses.

TOBACH: I want to get that from you later on because in terms of how animals see different kinds of colors in different seasons because of hormonal change is a very interesting problem which could be related to this.

DOSE: I don't think this process is related to the composi-tion of the lens.

TOBACH: It's something to look at.

RUTTER: The relationship of your talk to biological individ-uality is that life on Earth is one aspect, is an individual aspect, of a number of sets of possibilities. That is, a set of possibili-ties that would be based for example on different chiral forms of amino acids, sometimes carbohydrates, etc., or even positive matter-antimatter (we didn't talk about that). But still there is a dis-tinct possibility.

DOSE: There are constraints posed by the properties of matter in this Universe. We don't know anything about the origin of this matter or of the origin of the natural laws governing the evolution of elements. I see no chance for forms of life not based on carbon. But I could think of life based on D-amino acids or L-sugars.

RUTTER: Some of these elements of life are really selected presumably by their fitness.

DOSE: That is correct. I am glad to have this term, fit-ness, mentioned again. Already this morning we talked about the fitness of cells. We may transpose this term on to the fitness of molecules, even to the fitness of elements. Also, carbon nuclei, for example, show the fitness to "survive" in the interior of cer-tain stars.

RUTTER: And that's very well described by both Henderson and Wald, who developed that so extensively. Then there's another aspect of life which is purely stochastic in the sense that it could be one chiral form of amino acid or another chiral form of amino acid. And maybe that general principle then could be extended to the aspect of life which we are talking about; that is, that some things are fixed by nature of the process itself and other things are purely stochastic.

DOSE: Or these things presently appear stochastic to us.

FOX: It seems to me that the relationship to the question of individuality particularly resides in the phenomena that Klaus Dose was talking about--self-ordering of amino acids. If what we think is a correct model for how precursors of life and the first cells began, a key phenomenon in the whole sequence is the self-ordering of amino acids which is a _very_ self-limiting process.

KOSHLAND: I remember making a calculation once on the genetic code. In fact, in spite of what people say, if you just distribute the three-letter codes randomly, the damage of a mutation is almost the same as in the present code. So, it pretty well argues for the fact that once this code started to be established, it was so much better for organisms that it got established as the going mechanism. Organisms that are better adapted to the environment, and have a one percent edge over other organisms, will essentially take over completely in a few generations. A lot of codes would have been satisfactory. It depends upon which one got established first. So, I would think that prebiotic evolution could be just as arbitrary as post. . . .

FOX: So far as genetic code is concerned, you now are no longer talking about primordial events.

KOSHLAND: Oh no. We are now much later.

FOX: Those of us that adhere to a proteinoid-first point of view or a proteinoid-first bias if you wish, have extended that interpretation to the identical conclusion that you infer--that there could have been a large number of genetic codes and a selection was made from those (Nakashima and Fox, _Proc._ _Nat._ _Acad._ _Sci._ _69_, 106-108, 1972). That's the one that overgrew the population of possible genetic codes.

CAPRA: It's likely that any code would be superior to the nonrandom association of amino acids which must be an extremely inefficient process. Despite what you have shown.

FOX: Why? I can understand the assumption that genetic coding would be more efficient. But, random association would, I think, be more inefficient. Any one association could occur only once in a literal blue moon from a random matrix (Wigner, in "The Logic of Personal Knowledge," 1961, and Eden, "Mathematical Challenges to the Neo-Darwinian Interpretation of Evolution," 1967).

CAPRA: Well, what I meant to say was that you can put amino acids together and heat them and generate a few tripeptides, for example.   There's a certain predictability to that but any time you could superimpose upon that a nonrandom association that could then be selected for you would have an enormous advantage over the continual selection of the oft-assumed random association of amino acids.

FOX: To emphasize my point, the two tyrosine-containing tripeptides that Nakashima found by heating glutamic acid, glycine, and tyrosine (Nakashima et al., Int'l. J. Quantum Chem. Quant. Biol. Symp. 4, 65-72, 1977) and which Dose and his associates confirmed (Hartmann et al., BioSystems 13, 141-147, 1981) are the only tripeptides formed.   No others.   They dominate the product.

I have heard others comment on the fact that electropherograms of proteinoids (Dose and Zaki, Z. Naturforschung 26B, 144-148, 1971) directly suggest coded macromolecules.   The question of relative degree of nonrandomicity in self-ordered macromolecules vs. coded macromolecules is being approached.   Newer data on order in the Universe requires that we examine this question again (Fox, Naturwissenschaften 67, 576-581, 1980).

CAPRA: I wanted to make just one other point that ties the first two talks together.   And that is, and again I am sure there are billions of years of evolution between pyroglutamic acid, tyrosine, glycine, and bacteria but in modern neurochemistry and growth factor chemistry there are over half to three-quarters of the chemoattractants, the small neuropeptides, the growth stimulators, are relatively small tripeptides and most of them begin with pyroglutamic acid as their amino terminus.

FOX: Yes.   It is perhaps worth mentioning the history that reveals that most modern organismal pyroglutamate peptides, which are now quite a number, were first found after pyroglutamic acid was identified as a key intermediate and N-terminus in model prebiotic experiments.

TOBACH: In the selection that you're talking about, do you think that that was also random?

FOX: ,   Which selection do you mean?

TOBACH: Of those early proteinoids.

DOSE: Like what?

TOBACH: Was that a random process as to which ones got to do more things?

DOSE: I think it was a selective <u>non</u>random process. Individual proteinoids could have been separated on clays, for instance.

TOBACH: .   In other words, there was something else happening at
the time.

DOSE: There was self-organization and selection by interaction with the environment.

TOBACH: So that, therefore, when you talk about the selection
process, you are again talking about the connectiveness with a
relationship between the structures that exist and what's going on
around them. That's the question, you see, that when you talk
about random or nonrandom or semirandom and whether it's fixed or
determined, I think that one always has to put the process you're
looking at into the context. It's not taking place in vacuo.

FOX: I agree that the process of selection does not occur
in vacuo. Therefore, I don't telescope variation and selection
into the exact, identical time frame. In this way, I look at the
evolutionary process that was (a) variation and <u>then</u> there was
(b) selection from among the variants, unless, of course, you
ascribe creative power to "natural selection." The point that I
believe we are discussing here is whether the <u>variation</u> was random
or nonrandom. But in answer to your question, the first step was
the variation; the second step was selection from the products of
that variation, much as Darwin defined the relationship; and it
then has to be the variation that was either random or nonrandom.

TOBACH: Right. And the reason that they are not random is
because of what else was going on at the time the variation began
to be.

FOX: We believe that they are not random for internal
reasons. Molecules have different shapes, unlike playing cards
that have the same shape and thus participate in random arrange-
ments.

TOBACH: That part of the story will then determine what's
going to happen next.

, FOX: I think that's the fountainhead.

TOBACH: All right. But that is the point. In other words,
when we argue about determinism or randomness, we are trying to
make an either-or categorization which doesn't fit reality. Because
what's always happening is that things are both determined and
probabilistic or random at the same time. In other words, because
something is the way it is right now, what happens later is going
to be the consequence. If it were different from what it is now,

it would be different later on. And because the shape of these proteinoids have certain characteristics, they are more likely to
respond and to change in a given way depending on what else is happening. At the same time, what else is happening is not always
predictable because you have many lines of development going on at
the same time. And as they cross, they bring out something different each time. But you see, that's why we should not get into these
arguments about determined or random because they are both true in
a sense.

FOX: Let me develop the materialistic position a little
more. I believe you are arguing for a powerful influence of the
environment on the individual. My interpretation of the experiments
is that the variation is not determined randomly; it is largely
determined endogenously. After a nonrandom beginning, probabilism
or whatever else is superposed, but then the whole evolution is
nonrandom.

          I don't think it will finally be shown to be proba-
bilism, or due to the vagaries of the environment. I agree that
the environment has a powerful effect on the individual. The environment includes numerous other individuals. If the behavior of
all of the individuals is nonrandom and genetically predetermined,
their interactions would be, I believe, predetermined by some geo-
metrical function of their individual predeterminisms. But now the
behavior of the total group has generated complexity to a geometri-
cal extent. One cannot identify and analyze the processes anywhere
near as easily as he can for a single individual. To get to details
in this view, I believe the analysis, in effect, has to be a synthe-
sis--a constructionistic integration of what we can see most easily
in an individual.

TOBACH: That's what's so beautiful about that synthetic work,
because that's true for all of the way matter operates.

STANFORD: Just a quick philosophical comment and question. For
some time I have been intrigued by the tenacity with which scientists hold on to their views.

          In Thomas Kuhn's book, which just fascinated me, he
explains the vigor with which scientists hold to a widely accepted,
"establishment" paradigm. And I suppose that means that scientists
are beings after all with enormous egos. I notice, Klaus, in the
last part of your paper that you say that the theory of polyamino
acids-first is not widely held because it doesn't relate to contemporary molecular biology. I would infer then, bearing upon Dan
Koshland's paper, and keeping Kuhn in mind, that contemporary
molecular biologists look upon polyamino acids first as a toxic
gradient and swim away from it.

LAUGHTER . . . .

KOSHLAND: As a molecular biologist, I think you're being modest
and I think correct about what the problem is. It's hard to go back
to what really happened at that point because we keep erasing the
records. I think that it's as reasonable a hypothesis as any.
When we look for life on other planets, we are essentially starting
out with the idea that life has to be water-based; but, in fact,
ammonia is mentioned. As an organic chemist, you can run a lot of
reactions in liquid ammonia. It seems to me conceivable at least
that one could develop an NH₃ system and who knows that may be on
some planet . . . .

DOSE: .   There could have been a competition between a water-
based system and an ammonia-based system; the water-based system
would be fitter because water is more stable than ammonia.

KOSHLAND: Wouldn't it depend on the surroundings and the temperature? Below zero degrees, water would be frozen. Even a
silicone is very similar to carbon and . . . .

DOSE: Not so similar. You don't have the diversity in silicone compounds that you have in carbon compounds.

CAPRA: A quick question. What was found when they did amino
acid analysis on the lunar rocks? Did they find amino acids and
were they distributed in any way like the amino acids we see here?

FOX: The results that have been generally accepted, after
early challenge (S. Yuasa and J. Oró, in Science and Scientists,
pp. 31-37; Kageyama et al., eds., Japan Sc. Soc. Press, Tokyo,
1981) are from twelve collections from six missions in the Apollo
Program. Those gave upon extraction and hydrolysis, by a special
method that was developed for Apollo 11, five or six amino acids in
every locale of the Moon visited. Those amino acids were aspartic
acid, glutamic acid, glycine, alanine, serine and, in some cases,
threonine (S. W. Fox, K. Harada, and P. E. Hare, Subcell. Biochem.
8, 357-373, 1981). When others used the kind of method that were
developed for Apollo 11, they found that the meteorites that have a
reasonably hygienic history (none of them has a very clean history),
contain proteinous amino acids much like the lunar amino acids in
almost the same proportions. A lot more total amino acid was found
in the meteorites than in the lunar "fines", but when one equilibrates
these figures by dividing by the amount of carbon in each of those
two sources, the amounts are almost within the same range. Essentially, a very similar pattern of proteinous amino acids by kind and
amount was found in those two extraterrestrial sources.

CAPRA: Any dipeptides or tripeptides?

FOX: There is yet no way of knowing from the amounts that
were available for analysis whether amino acids from the lunar fines
were released from small peptides.

DOSE: I think you should make another point clear in this
relationship. There is no evidence that these amino acids are
present as such in lunar fines or in meteorites. The generally
accepted view is that these amino acids are formed from other precursors during the extraction of the material.

FOX: Your emphasis is well made; the amino acids are there
as precursors, which are hydrolyzed either upon extraction or separately to amino acids (Fox, Harada, and Hare, Subcell. Biochem. 8,
357-373, 1981). No one knows for sure what the precursors are;
they are believed to be cyanides. In fact, you can polymerize HCN,
hydrolyze it, and get the same six amino acids--or other sets
depending upon added minerals, etc. The hydrolysis to amino acids
occurs rapidly, which is a main reason for believing that the precursors are some cyanide polymer rather than di-, tri-, or higher
peptides.

MOLECULAR GENETICS AND INDIVIDUALITY

William J. Rutter

Department of Biochemistry & Biophysics, and
Hormone Research Laboratory
University of California
San Francisco, California 94143

INTRODUCTION

Individual humans and animals obviously can be distinguished
at the level of body form, physiology, biochemistry, and psychology.
The basis for this individuality must at least partially be genetic.
New advances in molecular genetics have illuminated the mechanisms
of regulation of genetic systems. This information tends to deempha-
size some simple models explaining the basis of individual variation
and suggests alternative hypotheses.

Mechanisms that Insure Genetic Variability

All organisms from bacteria to humans have mechanisms that
ensure genetic variation in individual members of the species.
During gametogenesis, individual chromosomes from the parental pair
are selected randomly in the process of meiosis to yield the set
present in gametes. Thus there are 2^{46} different possible combinations of chromosomes in a single offspring from human parents. The
variation is further enhanced by the stochastic exchange of homologous chromosomal segments between sister chromatids. It is estimated that about 10 sister chromatid exchanges may occur per
nucleus.

In addition to the biologically programmed means for generating genetic diversity, normal cellular activities result in chromosomal damage. Chromosomal breakage occurs in about 1% of the cells
and ring chromosomes may occur in roughly 1 out of 10^4 cells. Thus
on the average a human being (10^{13} cells) may have at least a
billion cells containing abnormal chromosomes. The functions of
these cells are usually covered by normal cells of the same type,

61

but if they occur early in embryological development, they could
result in an altered pattern of morphogenesis or function.

The genetic material is also subject to mutational events that
lead to changes in the DNA sequence or to additions and deletions
of nucleotide residues. These changes are a consequence of inherent
uncertainties in the replication process and to environmental vicis-
situdes involving the action of exogenous or endogenous mutagenic
agents. The level of adventitious mutation by radiation or chemical
mutagenesis obviously varies with the individual, the environment,
nutrition, etc. The mutational load is difficult to quantify, but
it is believed to be less than one per cell. Since gene coding
regions probably occupy less than 1% of the total nucleotide sequence,
most mutations would occur in neutral regions. These classical
mechanisms ensure that the genetic repertoire of each individual is
to some extent different, and they have historically provided a
plausible basis for the genetics of most individuals. They, however,
do not explain the uniqueness of monozygotic twins.

Jumping Genes: Mobile Genetic Elements and Altered Gene Structure

In the past it has been assumed that DNA, the genetic material
was like a static archival tape. Newer findings, however, indicate
that it is a more dynamic interactive structure. It is believed
that certain DNA segments have structural features which allow
transposition within the chromosome by natural processes present in
most if not all living cells. These transposable elements have
been studied extensively in bacteria, yeast, maize and Drosophila
but are believed to be present. They have not as yet been specifically demonstrated in human cells. The consequences of the mobility
of these genetic elements are not completely understood. If transposition occurs randomly throughout the genome as in the case of
bacterial transposable elements, then occasionally a foreign DNA
segment would intrude upon a functional sequence such as the coding
region of the gene; the resulting gene product would probably be
biologically inactive, so the cumulative effects would be negative.
However, it is also possible that the mobile elements contain structures that enhance expression of genes in the immediate vicinity of
the integrated DNA. This property is exhibited by the delta
sequences in yeast and by the so-called enhancer sequences present
in some mammalian viruses. Then, if the transposition occurred in
a somatic cell, the effects would be restricted to a single individual. The magnitude of the effect would depend on the time during
development that the mutation occurred. If the transposition
occurred in germline cells, the effects would be transmitted to the
progeny.

Certain viruses that integrate into the host genome exert
effects similar to transposable elements. Viruses such as Simian
Virus 40 or its relative polyoma, the oncogenic RNA viruses, and

hepatitis B virus are known to integrate essentially randomly into the DNA of susceptible cells. In most instances the integration would be in a neutral region (greater than 99%). If the viral sequences were inserted within the coding or regulatory sequences associated with the normal genetic repertoire the effects would most likely be negative. However the virus also contains regulatory elements, promoters and enhancers that may control expression of contiguous host DNA. Thus novel normally non-coded regions of the host DNA could be expressed.

It has also been demonstrated recently that exogenous DNA from any external source can be taken up by cells (though with variable efficiency). Over time the DNA can be incorporated into the genome. New genetic information can be introduced into the genome by transfection. Novel DNA sequences can also be generated from RNA species present in the cell by a process called reverse transcription that is catalyzed by an enzyme present in the RNA tumor viruses. Incorporation of DNA into the genome by either of these mechanisms, however, is considered to be a relatively rare event and thus not applicable to any general mechanism for explanation of individuality.

Hot Spots: Hypervariable Regions within the Chromosome

Most of the chromosomal DNA of the organism retains a constant sequence. It has recently been discovered, however, that there are regions of high variability interspersed throughout the chromosome. For example, my colleagues and I discovered and elucidated the structure of such a polymorphic region very near (365 nucleotides) the initiation of expression of the human insulin gene (Bell et al., 1981; Bell et al., 1982). This region was composed of tandem repeats of a family of oligonucleotides (14-15 bases in length) containing similar but not identical sequences. The variability in this region was due to both the number of oligonucleotide sequences in the tandem repeats, and also the arrangement of the oligonucleotides within the structure. The length of the region changes only slightly in a human generation so that DNA from genomes that contain the long region (i.e., 100 repeats) can easily be distinguished from DNA that contain a short polymorphic region (i.e., 20 repeats). The inheritance of long and short structures occurs in a Mendelian fashion; each behaves essentially as an allele. However, the sequences of variant oligonucleotides within the repeat are different for each individual. In fact, they are so strikingly different in the cases studied that differences may occur in somatic cells of the same individual. The polymorphic region may reflect some recombinative event or it may be a locus for integration of a mobile genetic element; it could also reflect some natural but as yet unknown chromosomal process including replication, segregation or determination of chromosome configuration. Thus the structure of this region could influence the expression of contiguous genes. The proximity to the insulin gene raises the question of whether

changes in the region could influence the expression of this gene.
A number of recent studies have correlated the length of the repeats
with a predilection to some type of diabetes.

Polymorphisms of other lengths have been detected in DNA. The
number of such hypervariable regions in the chromosome is not precisely known but may be 200-1,000 (Botstein et al., 1980). These
rapidly changing structures may help to determine or reflect the
unique characteristics of certain individuals. It seems unlikely,
however, that they play a general role in determining individual
variability.

The Deleterious Gene Hypothesis

The genetic burden of humans is frequently discussed. The concept implies that many alleles in the human genome are somehow
genetically compromised. The compromised gene is frequently viewed
as defective at the structural level. There is ample precedent for
this. The study of human genetics over the past 30 years has
revealed that many genetic lesions involve the production of abnormal proteins. The first example was the discovery that sickle cell
anemia is due to a mutation that causes a single change in an amino
acid residue in the protein (the replacement of a glutamic acid with
a valine in position 6 of the beta chain of hemoglobin). Subsequently
more than 300 different hemoglobin variants have been discovered
and their characteristics (some neutral, some deleterious) have been
defined. Further mutations in many other genes have similarly been
recognized. The evidence strongly suggests that the pattern of
mutations studied in hemoglobins (e.g., amino acid (nucleotide) substitutions, deletions or additions) is similar in all other genes.
Thus it can be proposed that individuality is the summation of the
genetic defects. Each individual is genetically compromised but in
different genes and to different degrees. Thus the mix of the
genetic defects determines individual characteristics. Although a
genetic defect undoubtedly contributes to and may be a dominant
characteristic of single individuals, I believe it is unlikely that
defects in structural genes per se are the major basis for individual variation in general. Structural variations such as those seen
in hemoglobin occur rarely in humans unless there is some environmental selection (such as there is for hemoglobin S) which enhances
the proportion of a particular mutation in the population. The
rarity of these structural mutations (approximately one per 100,000
in the population) and the ability of the normal alleles to compen-
sate for the deficiency mitigate their penetration.

Conservation of Gene Coding Regions

The four nucleotides present in DNA and the three-nucleotide
codons provide 64 (4^3) different coding choices. Three of these are
used as termination signals, and the remaining 61 codons are used to

code for the 20 amino acids found in proteins. This coding redundancy is distributed nonrandomly throughout the amino acids, e.g., methionine has a single codon while leucine has six different codons. The variation in the codons coding for a particular amino acid usually occurs in the third base (wobble) position. The consequence of this is that the coding region of the gene can vary considerably while still prescribing an identical amino acid sequence. Studies of amino acid sequence and individual proteins of humans and related primates show great conservation of sequence. On the average a change of a single amino acid in a protein sequence occurs in approximately one million years of evolution. This fact hardly corresponds with a hypothesis requiring great variability in the structural genes in individual human beings. The corresponding changes in nucleotide sequence have not been as extensively studied; but are obviously more frequent. Recent studies of cloned human insulin gene sequences, for example, suggest that there may be variation at a level of approximately one per thousand nucleotides (Bell et al., 1980; Ullrich et al., 1980). When these changes are in the coding region, they largely occur in third base positions, thus predicting proteins of identical amino acid sequence. The conservation of sequence is partly a result of the high fidelity of the DNA synthetic process and also well-developed mechanisms for the elimination of errors (DNA repair). In addition, it appears that there are additional mechanisms of rectification of DNA sequence which may depend upon an absolute correspondence of DNA sequence (such as required in DNA recombination). These mechanisms help to insure the fidelity of structure of coding sequences in the germline DNA.

Gene Splicing and Variability

It has been discovered in recent years that the coding regions of the genes are not continuous but contain intervening sequences (introns) that must be removed by splicing such that the coding sequences (exons) are brought into precise register. This occurs for the most part at the RNA level but can occur in certain restricted circumstances at the DNA level. In the immunoglobulin genes, different DNA segments are brought together to determine the various regions of the antibody molecule (Early et al., 1980). In the case of the RNA transcripts, the intervening sequences are excised and the exons joined with great accuracy; dropping a single base would change the reading frame of the coding sequence with the result that an aberrant amino acid sequence would be produced. The splicing is, however, very sensitive to both nucleotide sequence and the environment within the cells. Alternative splicing of the same gene occurs in normal and reconstituted systems (Young et al., 1981; Inana et al., 1983; Rosenfeld et al., 1983; Laub and Rutter, 1983). Nevertheless, during evolution many amino acid sequence changes, especially length polymorphisms, can be mapped to splice junctions. The migration of exon junctions along the gene (junctional sliding) may be a prominent mechanism that accounts for

these changes in sequence (Craik et al., 1983). The DNA joining
which occurs in the construction of functional immunoglobulin
genes is much less precise; indeed the variation in sequence occurring at particular junctions may contribute to the generation of
antibody diversity (Early et al., 1980).

Variability in Gene Expression

In the various cells of the higher organism the genes are
selectively expressed. Some of the genes involved in certain common
activities are expressed in all cells, others are expressed in a
certain subset of the cells, and still others are expressed in a
single cell type. For example, hemoglobin is expressed only in the
erythrocyte and insulin expressed only in the B cells of the islets
of Langerhans of the pancreas, etc. The ability to selectively
control gene expression in differentiated cells can be truly extraordinary. There is evidence suggesting that control magnitudes of
at least one million fold are involved. One of these levels of
control may operate at the level of chromosome structure. It is
well known that the chromosomes are present in complex folding
patterns and that the conformations may differ in various differentiated cells. Further, it has been demonstrated that the DNA in
the chromosome region of an expressed gene is more susceptible to
exogenously added degrading enzymes suggesting that the composition
of the chromosome is somehow different (Weintraub and Groudine,
1976).

The expression of the genes involves the synthesis of an RNA
transcript, splicing of the RNA and subsequently synthesis of the
protein. Like the splicing mechanism, transcription and protein
synthesis occur with a high degree of fidelity; thus the great
majority of proteins produced from a single gene is identical in
structure. However, there are also cellular mechanisms that detect
and destroy aberrant proteins and hence the homogeneity of the proteins that result from the gene is insured. The control of the
expression of the genes might occur at any one of the stages in the
formation of the protein but the transcription of the DNA, the
first step in the process, is believed to be the major control site.
Transcription is effected by a complex enzyme, RNA polymerase, which
is composed of about 10 polypeptide subunits. The initiation and
termination of transcription by the RNA polymerase are precisely
controlled by signals in the DNA, and by auxiliary proteins which
effect the entry and departure of the RNA polymerase from the DNA
template. The region of the DNA-controlling initiation of transcription is dependent upon the nucleotide sequence in a region of
the DNA upstream from the coding sequence. In higher organisms,
including humans, the promoter region is complex and is comprised
of at least three elements. The sequence of these elements can be
quite variable from gene to gene, and the rules defining the
sequence(s) requisite for efficient expression are as yet imperfectly

understood. Single nucleotide changes within the promoter region
exert variable effects on the efficiency of expression, depending
apparently on their specific involvement in the binding of the
polymerase or auxiliary proteins required for transcription. In
addition to these core promoter regions there are other sequences
which can dramatically affect transcription. A positive regulatory
element termed an enhancer, has been discovered in a number of
viruses (Benoist and Chambon, 1981). It has recently been shown
that sequences just outside the core promoter are involved in
selective expression (Walker, Edlund, Bonlet and Rutter, submitted;
Gillies et al., 1983; Kondoh et al., 1983). Some of these specific
control elements are located upstream from the core promoter (Walker
et al., submitted). Enhancer-like sequences located in the intron
regions have also recently been implicated in the specific expression of the immunoglobulin genes (Gillies et al., 1983).

This brief review indicates that the regulation of expression
of specific genes is extremely complicated, and is not yet fully
elucidated, but it depends upon the DNA sequence in several regions
as well as interactions with several classes of proteins. Further,
individual mutations can affect rates of transcription substantially.
The regions of nucleotides involved in the regulatory regions so far
discovered are 200-300 bases and the number of genes in the human
haploid set is approximately 25,000. Thus, the composite regulatory
domains of an individual genome may approximate 10^7 nucleotides. If
mutations occur at a rate of 1 in 10^6, then the regulatory regions
of 10 genes per individual cell may be affected. The human is comprised of approximately 10^{13} somatic cells. Thus the probability
that some of these cells contain mutations which positively or nega-
tively affect the expression of some subset of genes is very high.
Of course the effects are mitigated by diploidy of the chromosomes;
the probability of affecting both gene copies associated with a
single function is much more modest. Nevertheless the net rate of
expression is a summation of the rates of expression from both gene
copies. For certain genes the level of expression is critical.
There is abundant evidence of profound gene dosage effects (the
effects of different rates of transcription due to different gene
copy numbers). For example, chromosome 21 trisomy (three instead
of the normal two chromosome 21's are present) results in Down's
Syndrome (Mongolism). This complex condition involves typical
developmental changes resulting in changes in stature, body conformation and mental acuity. Presumably this condition results from
the enhanced expression (50%) from one or more genes present on
chromosome 21. Alterations by mutation in the regulatory regions
of the gene could in a similar fashion produce changes in the rate
of expression of that gene such that the consequences would be
similar to a gene dosage effect. The probability that such changes
would be physiologically relevant would depend upon (1) the specific gene involved (only certain key genes that control an
important functional system would be likely to exert a profound

effect), and (2) the period in development at which the mutation
occurred. If the mutation occurred at a relatively early stage in
the developmental process, all of the progeny cells would be
affected and hence a significant fraction of the total tissue could
be involved. This subtle modulation of rates of expression may be
responsible for some of the differences in human beings of essentially common genetic constitution (for example, monozygotic twins).

Variation Caused by Different Levels of Regulatory Molecules

The steady state level of any compound is obviously determined
by the net rate of formation minus the rate of disappearance of the
compound. In some cases, the absolute steady state concentrations
of a gene product may vary over wide ranges with little physiological consequence. For example, above a certain critical level,
enzymes catalyzing one step in a metabolic pathway can be present
at quite different concentrations without affecting the total flux
in the pathway. The net result would be alteration of the concentration of the intermediates. On the other hand, the steady state
concentration of regulatory molecules such as hormones that control
a particular biological function is critically important. The
functions of the major organs and tissues of the body are controlled
by one or more hormones. The major hormones such as insulin, growth
hormone, the sex hormones, thyroid hormone, etc., are well known.
It is less generally recognized that there is a large number of molecules (usually termed factors) that resemble classical hormones in
their mechanism of action and that affect growth and function of
virtually all cells. The number of these developmental hormones
and the range of their activities is not yet completely known.
Indeed, new factors are frequently being discovered. However, it
is believed that the major aspects of development are regulated by
these developmental hormones. The regulation of the synthesis of
these hormones is complex. For example, the endocrine cell must
somehow serve the need for the hormone by monitoring some compound
that reflects this need. Small changes in the rate of production
or the rate of disappearance of these compounds can have a relatively large biological effect. We expect that the synthesis of
the hormones by an endocrine cell is dependent on intracellular
molecules interacting with the gene and on other effectors interacting with the cell. Thus the synthetic rate may vary in individual cells around some Gaussian mean.

The biological response to protein hormones is due to an interaction with receptors present in the membranes of the target cells
and to a series of reactions coupling this receptor/hormone interaction with the regulated product or process. The numbers of receptors on the cell membrane and their affinity for the hormone as
well as the efficiency of the coupling to the biological effect will
also to some extent be variable. Thus we expect individual cells
to vary somewhat in their capacity to respond to the hormone around

some Gaussian mean. In multicellular organisms like humans a second order of complexity is present. The net production of the hormone is the summation of the production from all cells. Thus the regulation of the cell number becomes crucial. If the regulatory mechanisms tolerate a difference of one cell division, a net difference of twofold in the cells producing the hormone would result. Further, the concentration of the hormone in the serum will depend on the rate of secretion as well as the rate of disappearance through turnover (largely occurring in the target cells responding to the hormone). Finally, the number of target cells is also controlled by some set of factors and will vary around some Gaussian mean. Thus the magnitude of the biological response and also the rate of destruction of the hormone will be related to the number and characteristics of the target cells. In practice it is found that in normal individuals the hormone molecules themselves remain identical in structure, but the level of the hormones varies considerably. The difference in hormone levels among individuals is no doubt in part a result of genetic influences on the many factors controlling those levels. In a common genetic background such as monozygotic twins the variation is dramatically reduced; however, the levels are still not identical. These gratuitous variations might also be the result in certain cases of mutations in somatic cells that affect some part of the process.

Experientially-Induced Variation

In addition to genetic variation and the imprecision of regulation in a constant genetic background other factors can have a profound influence on the development of an organism and its subsequent biological characteristics. For example, the use of the optical system during a period of development can profoundly influence the connections made by optic nerves in the brain. Thus an eye which is artificially covered to preclude visual experience becomes ineffectually innervated (Wiesel and Hubel, 1965). For a certain period of development these effects are reversible, but subsequently the effects become irreversible and the eye lacks function. The degree to which this experiential induction of development applies to other regions of the brain or to other organs is not known. There are prolonged consequences of nutrition, for example, the number of adipocytes (fat cells) are apparently determined at a relatively young age and are influenced greatly by the amount of lipid in the diet. Thus it appears that the biological and physiological characteristics of individuals can also be influenced by nongenetically programmed variability.

SUMMARY

There are multiple bases for individuality, some of them strictly genetic, others are epiphenomenal. They include:

1. Stochastic chromosomal selection, and sister chromatid exchange which provides an individual genetic repertoire composed of normal and partially functional or nonfunctional (deleterious) genes.

2. Further genetic variability caused by mutations in somatic cells. These can affect both the structural (coding) regions of the genes and the regulatory regions. Some of the mutations in the regulatory regions should alter significantly the efficiency of expression of genes coding for hormones or differentiation factors.

3. Slight variations in the rates of synthesis of crucial regulatory molecules (hormones and differentiation factors) caused by mutation or by stochastic variation within the limits of the natural regulatory system may result in significant changes in relative growth rates of particular regions of the body (morphogenesis) and of general physiological or mental characteristics.

4. The characteristics of the adult organism may also be influenced by nutritional status and sensual experiences during the developing years.

REFERENCES

Bell, G. I., Karam, J. H., and Rutter, W. J. 1979. Polymorphic DNA region adjacent to the 5'end of the human insulin gene. Proc. Nat. Acad. Sci. U.S.A. 78, 5759-5763.

Bell, G. I., Pictet, R. L., Rutter, W. J., Cordell, B., Tischer, E., and Goodman, H. M. 1980. Sequence of the human insulin gene. Nature 284, 26-32.

Bell, G. I., Selby, M. J., and Rutter, W. J. 1982. The highly polymorphic region near the human insulin gene is composed of simple tandemly repeating sequences. Nature 295, 31-35.

Benoist, C., and Chambon, P. 1981. In vivo sequence requirements of the SV 40 early promoter region. Nature 290, 304-310.

Botstein, D., White, R. L., Skolnick, M., and Davis, R. W. 1980. Construction of a genetic linkage map in man using restriction fragment length polymorphisms. Am. J. Hum. Genet. 32, 314-341.

Craik, C. S., Rutter, W. J., and Fletterick, R. 1983. Splice junctions: association with variation in protein structure. Science 220, 1125-1129.

Early, P., Huang, H., Davis, M., Calame, K., and Hood, L. 1980. An immunoglobulin heavy chain variable region gene is generated from three segments of DNA: V_H, D and J_H. Cell 19, 981-992.

Gillies, S. D., Morrison, S. L., Oi, V. T., and Tonegawa, S. 1983. A tissue-specific transcription enhancer element is located in the major intron of a rearranged immunoglobulin heavy chain gene. Cell 33, 717-728.

Inana, G., Piatigorsky, J., Norman, B., Slingsby, E., and Blundell, T. 1983. Gene and protein structure of a β-crystallin polypeptide in murine lens: Relationship of exons and structural motifs. Nature 302, 310-315.

Kondoh, H., Yasuda, K., and Okada, T. S. 1983. Tissue-specific expression of a cloned chick δ-crystallin gene in mouse cells. Nature 301, 440-442.

Laud, O., and Rutter, W. J. 1983. Expression of the human insulin gene and cDNA in a heterologous mammalian system. J. Biol. Chem. 258, 6043-6050.

Rosenfeld, M. G., Mermod, J-J., Amara, S. G., Swanson, L. W., Sawchenko, P. E., Rivier, J., Vale, W. W., and Evans, R. M. 1983. Production of a novel neuropeptide encoded by the calcitonin gene via tissue-specific RNA processing. Nature 304, 129-135.

Ullrich, A., Dull, T. J., Gray, A., Brosius, J., and Sures, I. 1980. Genetic variation in the human insulin gene. Science 209, 612-615.

Weintraub, H., and Groudine, M. 1976. Chromosomal subunits in active genes have an altered conformation. Science 193, 848-856.

Wiesel, T. N., and Hubel, D. H. 1965. Comparison of the effects of unilateral and bilateral eye closure on cortical unit responses in kittens. J. Neurophysiol. 28, 1029-1040.

Young, R. A., Hagenbuechle, O., and Schibler, U. 1981. A single mouse α-amylase gene specifies two different tissue-specific mRNAs. 1981. Cell 23, 451-458.

DISCUSSION

KOSHLAND: A consequence of your hypothesis is that those individuals that you found had polymorphisms should have less insulin
in identical cells. Do they?

RUTTER: It isn't a consequence of our hypothesis. We're not
sure that those sequences have anything to do with insulin regulation. In fact, we don't think they do.

KOSHLAND: So those polymorphisms are not in the regulatory
region.

RUTTER: They are outside the regulatory region, albeit only
a few nucleotides. The reason why we don't think that they are
regulatory for human beings is that there is not a similar regulatory region for rats. And, we are just now going down to the
primate level in our studies. We just don't know what those
sequences are. My own hunch is that those sequences are part of
a cassette which is moving a great huge piece of DNA around in
human beings.

KOSHLAND: So your hypothesis at the moment is that the polymorphisms probably are occurring in regulatory regions. It could
be, then, that the regulatory regions are as carefully selected as
the gene itself and we can't tolerate those changes.

RUTTER: It certainly could be. If, for example, one examined
insulin genes or the nucleotide sequence between human and rat and
between human and fish insulin genes, one finds that the regions
bear a good deal of similarity but not as much similarity as in the
coding regions. So, the postulate I make is that very subtle
changes in the control regions are possible. This follows from the
fact that one can't recognize a consensus sequence. One always can
locate a sequence in the coding region, but one can't recognize
relevant sequences in these control regions. So, there must be a
greater variability tolerated in these sequences. The net consequence is subtle variations in control, rather than dramatic regulations. It's not on-and-off, it's a hundred to seventy-five or
a hundred to ninety-five, something like that.

KOSHLAND: Do different organs of the same animal have polymorphisms, that is if you get the genes out of liver versus kidney?

RUTTER: We have been unable to perform that experiment as
yet; it's technically difficult.

KOSHLAND: It's easier to ask questions than to do experiments.

RUTTER: Certainly it's a major question.

CAPRA: Did I understand you to say you sequenced the human insulin gene from five thousand different individuals or was that done by restriction mapping?

RUTTER: No, the genes of two individuals were sequenced. We have sequenced five thousand bases in those two individuals. And it's in the five thousand bases that there were really only four nucleotide replacements.

CAPRA: There's just one other comment. I suppose, having described all those beautiful regulatory processes, everyone of us can think of another regulatory process that you haven't mentioned. But, there's one in immunology that fits beautifully to what you have described and I am sure you are familiar with, that may be of interest to the others, and that is the antibody molecule is both a receptor and it's secreted out of the cell. And one of the clever techniques that is used by immunoglobulin-secreting cells is that the very same molecule does both these things and the only difference is that long messenger RNA that Bill put on the board has two endings. And, you use one ending if you want it to be put in the membrane as a receptor, and then you use a second ending of the molecule if you want it to be secreted. So, it's such an extraordinary economy on the part of the cell to use effectively the same messenger, and it can therefore regulate how much the cell is bombarded by antigen and stimulated, versus how much of the molecule it is making it is putting on the surface versus how much it's secreting by effectively changing the end of the story by making this, and it's done by differential splicing.

          I would like to just extend one thing you said, Bill, and not completely give the talk I had intended to give tomorrow. These repetitive sequences that you describe that are in the noncoding region of the insulin gene are remarkably similar to, again, a phenomenon in immunology that is essential for the variability from molecule to molecule and that is the so-called D-region of the immunoglobulin molecule which is the real business part of the antibody molecule that dictates specificity. And there we see that every molecule can be longer or shorter and it can have these concatenated sequences of repeating units of gly-lys, gly-lys, gly-lys, or gly-tyr, gly-tyr, gly-tyr, and when you look in the DNA, it seems to be made up of these same sorts of building blocks. There, that's essential to variability. I think the extension in immunology of what you're talking about is that, and this is really the core of what I wanted to talk about tomorrow; every cell--the product of every cell--is different from the product of every other cell, possibly by some of these very same mechanisms.

HARDIN: I am going to ask a very simple-minded question. Forty or fifty years ago the question was asked, considering most mutations are rare in an individual: Can evolution, in fact,

proceed? Can there be selection for the process of mutation itself,
selection for high rates of mutation? We just couldn't see any way
in which the rate of mutation could be subject to selection. I
believe A. H. Sturtevant was the first to point this out. Listening
to what you've been saying, I wonder if the earlier conclusion needs
correction. Is this system that you have been describing one that
really should be looked upon as a marvelous system of avoiding variation in the regulatory system? Is it a new way of conservation?

RUTTER: I can make a couple of comments about the conservation
mechanisms, now called gene conversion. It's a way of allowing a
gene to monitor its structure via a neighboring gene so if this is
done frequently enough, they will all turn out to be the same. And
it turns out that it's known in the case of the globulin genes and
thought to be very much associated with a monotonous sequence of
$(GT)_n$ structures. In a whole series of genes of the pancreas, we
see this kind of structure. The observation initially made by
Smithies was that, when you find a sequence like that in the globin
genes, everything to the right-hand side of that has very little
variation. Since this could be the basis of the beginning of
reciprocal recombination, he thought this was a signal for the
"conversion." Regardless of whether that's a specific mechanism or
not, the fact is that you get enormous conservation in these structures.

          We can also speak about variations. We have found
this region in the insulin gene where it's hypervariable. These
polymorphic sites have been used by White and Botstein to map
regions of the DNA; there are roughly two hundred to a thousand
polymorphic regions per genome. This is not one per gene. They
are found in a number of organisms and they are not the same set;
they are slightly different, so far mostly variations of the one
we discovered. There is another kind of variation, one in which
the DNA is partially double-stranded. We know that in viruses, in
fact, the hepatitis B virus is partially single-stranded. In the
single-stranded region, one may get an enhanced sequence of variability because there's no way for it to rectify itself by using
the complementary strand. That may be the reason why there are
many strains of hepatitis B virus. Similarly, in influenza viruses
certain regions are highly variable. This allows the virus to
elude antibodies. So, there is a selection for hypervariability.
Capra's point of the usefulness of this variation is demonstrated
in these systems as well.

DOSE: I want to take up the issue of mobile genetic elements. What's known about the control of this type of gene and
what is the advantage or disadvantage of this process?

RUTTER: In bacteria, these mobile genetic elements frequently
contain a gene or a set of genes which confer some very important

biological function on the cell that harbors them, for example, resistance toward some antibiotics. They move around in the genome of some organisms or else they become part of a little plasmid (minichromosome) which can be used to infect different cells. The process is stochastic; the rate depends on the organism but there's no specific control.

DOSE: There's not!

RUTTER: In yeast, there's the so-called delta regions.

DOSE: The yeast nucleus or the yeast mitochondria?

RUTTER: These are in the nucleus and these move around so rapidly that if you take a culture in your own laboratory and let it grow for a little while and then restart it again, the structure of the DNA will vary because delta sequences are moving around wildly. The interesting thing about this delta sequence is that it has been shown in one case (the histidine gene) to dramatically affect the expression of the gene, even when it exists about 2,000 bases upstream from the gene. The best-described situation is the mating type in yeast.

CELL INDIVIDUALITY AND CONNECTIVITY, AN EVOLUTIONARY COMPROMISE

Werner R. Loewenstein

University of Miami School of Medicine
Department of Physiology and Biophysics
P.O. Box 016430
Miami, Florida 33101

I deal in this essay with individuality at the cellular level. This is a far cry from individuality of a whole organism, but it is something a reductionist can try his hand at in the hope that eventually the roots of that which sets one organism apart from another (similar) one may be traced to its cells, to their molecular machineries and interactions with each other and the environment. I try to look for clues at the interaction level.

The categorical statement of individuality was made in 1838 by the founder of cell theory, Mathias Schleiden (1838): "The cell is a circumscribed autonomous unit." This absolute statement goes far beyond what we would now find necessary for characterizing individuality of a cell or of any other entity. In fact, for all their extraordinary fruitfulness in propelling biology over these years, we know now the notions of cellular circumscription and autonomy to be wrong. It is instructive to see for the present purposes, why they were wrong.

First, there are thermodynamic reasons why a cell or, indeed, any living system made of nonsolid-state ephemeral materials could not be autonomous (Schleiden's postulate anteceded the development of thermodynamics). The entropy in such a system increases so rapidly that left on its own it would reach deathly equilibrium, probably within hours, certainly within days. To avoid such premature decay, the cell takes in negative entropy from its surrounds. It took aeons for the cells to perfect this entropic extraction-- a supreme magician's trick that comes as close as anything we know to life's essence. As so often when real wonder stares into our eyes, we escape into studying tangential things, like feeding and metabolism, never to see the forest for the trees. But whether

77

we look at the extraction process in its thermodynamic essence or
merely at the tangential material exchange, it is clear that there
cannot be autonomous cell circumscription.

A second reason for nonautonomy arose as soon as cells clustered to form organisms. Orchestration of the multicell activities
then became essential, and the selection pressures began to shape
devices for intercellular communication. The evolution of these
devices spans the entire phylogeny (and for our brain's sake we
hope it is continuing!), each new cell coordination requiring a
communication instrument. Some of the instruments are used over
and over for different coordinations--a common evolutionary parsi-
mony. Each new coordination achieved meant one more loss of cell
autonomy, and so among the 10^{15} cells in man there is hardly one
that is not interdependent with many others. What finally emerged
is a crisscrossing webwork of cells that have given up their
autonomy and often also, in a literal sense, their circumscription.
But they have retained their genetic uniqueness. It is that
uniqueness that one may call cell individuality, and I am inclined
to view the evolution of cellular communication systems as an endless maneuvering for achieving cell orchestration without loss of
such cell individuality.

The basic evolutionary strategy was to employ molecules as
communication signals: hydrophilic molecules using as vehicles
the aqueous phases of cell and intercellular space or hydrophobic
ones, membrane-bound, for the more short-range signaling. The
problem then was how to bypass the rather impervious cell surface
membrane of unicell times without disrupting it as a barrier
between cell interior and exterior. At first glance this may seem
like having one's cake and eating it too, but the riddle eventually
met with three solutions.

One solution involved a wolf-in-a-sheepskin ruse to get hydro-
drophilic signal molecules across the hydrophobic surface membrane.
The solution was to envelop the cytoplasmic signal molecules in
membrane vesicles that fuse with the surface membrane on contact,
releasing the signals to the external liquid, their final carrier
to the receptor targets on other cells (Fig. 1). This constitutes
the well-known <u>humoral</u> form of communication, which nearly all
cells engage in at one time or another.

Another solution was to imbed signal molecules into the surface membrane and let them interact with complementing molecules
on the membranes of other cells, directly or by way of a helper
molecule (Fig. 1). Known interactions of this sort are between
protein and glycoprotein molecules; one suspects that they are
means by which cells may say hello to each other in recognition.

Finally, a third way of bypassing the membrane was to tunnel

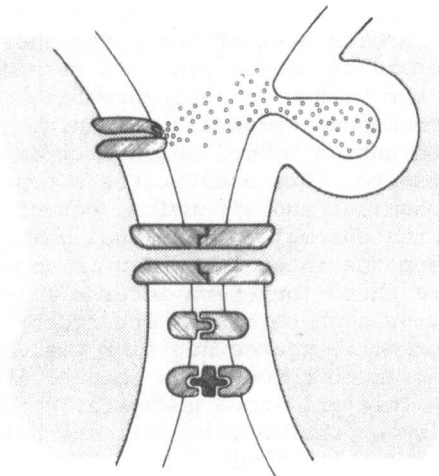

Fig. 1.   The three modes of intercellular communication.   See text.

through it, directly interconnecting the interiors of adjacent
cells. The tunnel evolved is an aqueous channel, about 20 Å in
diameter, made of two tightly joined protein halves, one contributed by each cell membrane (Fig. 1). This was another elegant
solution that at once preserved the continuity of the cell surface
diffusion barrier and the cells' genetic individuality. The preservation of the barrier was the result of a selection of a rather
leakproof abutment of the channel halves (and channel walls) and
the preservation of individuality, the result of a critical channel
bore size.

Of the three communication modes, that by cell-to-cell channels
involved the greatest loss of cell autonomy. In humoral intercellular communication there is a good deal of selectivity engendered by
the specificity in the signal-receptor molecular interactions, and
this also is probably so in the mode of communication by direct
interactions between surface membrane molecules. However, in cell-
to-cell channel communication there is actual loss of (individual)
cell circumscription, and the only selectivities we know are given
by the bore size of the channel and the polar effects of its fixed
or induced charges--the selectivities of a 20-Å sieve electrically
charged. Thus the macromolecules are stopped, but there is not much
hindrance for the cell-to-cell movement of molecules up to 12 Å
wide, and even neutral ones up to 20 Å wide (Loewenstein, 1981).

The gain for cellular communication from this evolutionary
compromise between cell connectivity and individuality was a

relatively free intercellular exchange of inorganic ions, metabolites, nucleotides, cyclic nucleotides, high energy phosphates, vitamins, small steroid hormones, etc. In respect to those cytoplasmic molecules, the connected cell ensemble, and not the single cell, became the functional compartmental unit. The immediate consequence was a coordination toward uniformity in this multicellular unit. In many instances, this constitutes a true Gleischschaltung, an evening out of chemical and electrical potentials in the cell ensemble; and in large ensembles individual variations are rapidly buffered. Only where the interconnection is compensated by energized transport, are there long-term intercellular gradients. (The known dynamic equilibria of this kind are fed by energized nonjunctional membrane transport--source and sink regions in transporting tissues; permeant movement through the channel itself, as far as we know, is passive). This primitive homeostatic role was probably the earliest function of the cell-to-cell channel, one that meant the greatest loss of cell individuality. Other functions were added later, adapted to the cell-to-cell dissemination of signal molecules regulating somatic and genetic processes, and among these are electrical signals controlling nerve, heart- and smooth-muscle activity and presumably signal molecules controlling the growth and differentiation of cells in embryonic and adult tissues.

Although the development of the cell-to-cell channel meant loss of circumscription for the individual cell, circumscription was preserved for the connected cell ensemble as a whole, the channel is rather leakproof to the exterior; thus the ensemble has a continuous carapace made of the channel walls and the cell surface membranes. This conversion from individual cell circumscription to multicell circumscription brought several gains for cellular communication. First, it allowed an intercellular signal transmission well insulated from the external forces of the environment. This probably provided a suitable milieu for the fashioning of auto-deterministic processes, as in embryonic development, that shape themselves by forces emanating from within. Second, it permitted a signaling by diffusible molecules within a medium of finite volume, namely the volume sum of the connected cells. This favored development of signal systems where information is simply and precisely coded into (intracellular) molecular concentrations (Loewenstein, 1979). Neither the humoral nor the surface-molecular interaction mode of cellular communication offered these advantages.

However, systems with multicell compartmental units are more vulnerable than systems in which each cell is such a unit; even one single damaged, leaky cell can endanger the survival of all other cells in the ensemble. Thus, mechanisms became necessary that permitted the damaged cells to be uncoupled. What evolved was a rapid closure mechanism of the cell-to-cell channel in which Ca^{2+} is a controlling agent. This channel closure ensues whenever a cell springs a leak or the intracellular Ca^{2+} concentration rises otherwise sufficiently above its low normal level (several orders lower than that in the cell exterior)--an elegant self-sealing mechanism of the carapace (Loewenstein and Rose, 1979).

Aside from controlling the channel open state, the size of the compartmental unit can be changed by regulation of the channel formation process. The lifetime of a channel is short (probably of the order of 10^1 hr) relative to that of a cell. Thus, by varying the rate of channel formation, connectivity can be regulated up or down (including total disconnection). In this regulation cyclic AMP appears to play a key role. It is much slower than the control by channel closure; up- or down-regulation of connectivity takes several hours. It is called into play during hormonal stimulation of adult tissues and, we suspect, also in embryonic development for (programmed) connection and disconnections switching molecular signals from one cell region to another (Flagg-Newton et al., 1981). The possibility of channeling signals by simply changing the extent of compartmental units makes this an extraordinarily versatile means for generating spatial and temporal patterns of cellular physiological states and differentiations.

There is a further remarkable aspect to this connectivity. The channel permselectivity can be varied, at least experimentally. In certain conditions, for example, the channel selectivity increases so as to effectively exclude the larger and charged (normally permeant) molecules from passing, while the smaller and less charged molecules continue to pass. The potential here is not only for a selective contraction of compartmental units to single-cell size, but given the right spatial controls, also for more modest contractions (Rose et al., 1977). Moreover, in certain conditions of cellular asymmetry, the channel exhibits a directional permselectivity; over a range of concentration, the (downhill) flux of large and charged permeants is very much greater in one direction than in the other (Flagg-Newton and Loewenstein, 1980). This adds a potential for directional filtering. It remains to be shown whether these potentials were actually realized. But a cellular communication system where signals can be transmitted differentially by merely changing some channel permselective properties is such a miser's dream of genetic economy for cellular differentiation that it is hard to imagine that evolution would have overlooked these possibilities.

In sum, the cells gave up their individual circumscription in respect to many diffusible molecules in the course of evolution but they retained it in respect to their DNA, RNA, enzymes and other proteins. This, then, is the sense of cell individuality: an apartheid for intracellular macromolecules and for supramolecular structures. Smaller intracellular molecules mingle with those of other cells influencing their somatic and genetic activities and certain extracellular molecules exert such influence less directly in addition. Modified this way, Schleiden's tenet of cell individuality loses none of the force it had for over a century.

REFERENCES

Flagg-Newton, J. L., Dahl, G., and Loewenstein, W. R., 1981, Cell junction and cyclic AMP: I. Upregulation of junctional membrane permeability and junctional membrane particles by administration of cyclic nucleotide or phosphodiesterase inhibitor. J. Membrane Biol. 63, 105-121.

Flagg-Newton, J. L., and Loewenstein, W. R., 1980, Asymmetrically permeable channels in cell junction. Science 207, 771-773.

Loewenstein, W. R., 1979, Junctional intercellular communication and the control of growth. Biochim. Biophys. Acta Cancer Rev. 560, 1-65.

Loewenstein, W. R., 1981, Junctional intercellular communication: The cell-to-cell membrane channel. Physiol. Rev. 61, 829-913.

Loewenstein, W. R., and Rose, B., 1978, Calcium in (junctional) intercellular communication and a thought on its behavior in intracellular communication. Ann. N.Y. Acad. Sci. 307, 285-307.

Rose, B., Simpson, I., and Loewenstein, W. R., 1977, Calcium ion produces graded changes in permeability of membrane channels in cell junction. Nature 267, 625-627.

Schleiden, M. J., Beitrage zur Phytogenesis, 1838. Muller's Arch. Anat. Physiol. Wiss. Medic. 1838, 137-176.

DISCUSSION

STANFORD: One of the greatest challenges to be emphasized
is to develop an appropriate balance between freedom and restraint.
The ethologists have referred to certain animal groups as maintaining successful continuity because they encourage the utmost expression of individuality, but not up the point where such expression
endangers the preservation of the group. You seem to be saying
something like this with respect to the cell, are you not, when
you describe the evolution of cellular communication systems as the
endless maneuvering for achieving cell orchestrations without loss
of such cell individuality?

LOEWENSTEIN: Yes. The comparison is apt. It was an evolutionary compromise cells made when they decided to make the jump to
form multicell organisms--a compromise somewhat of the sort individuals must make when they form societies. Organisms, really,
are in many ways like societies. The term organism is of recent
vintage. It was introduced not quite a hundred years ago by the
Count de Buffon, a French naturalist, as one began to realize that
animals were made of parts that were every bit as organized as
societies, where the various individual members interact with each
other. Basically, there are only three forms of interaction. One
of these I emphasized today to show my finder's bias. It also is
the one that endangered most the individuality of the cells in
organismic evolution and, therefore, required the greatest compromise.

STANFORD: Coming back to the word orchestration for just a
moment, what do you mean by orchestration?

LOEWENSTEIN: By orchestration, I mean just playing together.
Orchestra-playing together in concert, working in concert.

MONROY: I am wondering whether the orchestration you are
talking about which concerns individuality, applies to fusion of
gametes, sperm and egg. I think you cannot fuse sperm with sperm
and eggs with eggs.

LOEWENSTEIN: You are quite right. There is no evidence for
this kind of communication in gametes, nor in red or white blood
cells and, as I already said, in skeletal muscle and most nerve
cells. These cells probably underwent a secondary evolution to
prevent them from making the cell-to-cell channels. In the case
of skeletal muscle and nerve, the evolutionary objective was to
get rid of electrical cross talk between cells. These nerve and
muscle cells are head-strong individualists; each plays its own
fiddle, so to speak. The thousands of nerve fibers that make up
a nerve cable going to a muscle or coming from a sense organ are
generally well insulated from each other. They rarely make communicating junctions.

MONROY: Which may have been gap junctions.

LOEWENSTEIN: Well, "gap junction" is only a morphological term.
It stands for an electronmicroscopically recognizable entity where
presumably cell-to-cell channels occur clustered. I intentionally
avoided this term today, as I concerned myself with functional
aspects only. To add one point which may interest you; at any early
stage of embryonic development, even skeletal muscle and nerve cells
make cell-to-cell channels, communicating junctions. In fact, early
on, the entire embryo, all cells in the embryo, are interconnected
by such channels.

TOBACH: What you're saying about cells then would also be
true of, for example, the slime mold that you talked about earlier?

LOEWENSTEIN: Are you referring to the cell-to-cell channels?

TOBACH: I am asking about the channels that you talk about.
This is an entirely new concept to me. These are structural modifications that take place within a membrane. Is that right?

LOEWENSTEIN: Yes. They are protein channels.

TOBACH: Okay. Now, I'm getting to the molecular level.

LOEWENSTEIN: Just to answer your question about slime molds.
We don't know whether they have these channels. They certainly
do signal; they communicate by humoral signals. As Gerisch and
others have shown, they emit cyclic AMP as a signal and have receptors for that signal. Whether they have cell-to-cell channels in
addition, these slime molds, we don't know. Some years ago we
tried to find this out. Jamakosmanovic, one of my colleagues,
spent a good deal of time--not very fruitfully, I am sorry to say--
trying to see whether slime mold cells were electrically coupled.
To this end, one puts a pair of microelectrodes into neighboring
cells. But these cells were too clever; when pinched by the
microelectrode, they ran away. And my colleague was never able to
talk them out of that impish behavior.

TOBACH: The reason I asked you is that in your introductory remarks, you said something about when cells come together.
Unfortunately, to my way of thinking when cells proliferate, rather
as you say, are brought together, it's as though there were cells
floating around, and then something brings them together. However,
tissues and embryological development to me is always that in which
something keeps splitting and making more cells rather than cells
coming together. Then I thought about slime molds as something
that's coming together and if it's a question about coming together
that is involved in the channeling, as opposed to something that
develops from the cell, it's a very different process. I'm trying

to see the evolutionary picture. How many cells do we know in evolution that "come together."

LOEWENSTEIN: Well, the embryologists present here could tell
you better than I. There is a lot of migration of cells in any
embryo, active and passive movement of cells. Even in the primitive development you like, the slime mold, cells do actively shuffle
about; the running-away from our microelectrodes that I just mentioned is just a manifestation of the active ameboid movements of
these cells. To come back to real embryos, cellular movements, in
cell-dimension scale, can be over long distances. So cells "coming
together" is not at all a far-fetched situation. But I can make my
statement more general; whenever the cell membranes of channel-
competent cells get within 20 Angstroms of each other, they will
tend to make cell-to-cell channels.

TOBACH: But not with every cell.

LOEWENSTEIN: Yes. Cells are surprisingly promiscuous in this
regard. All cells in the early embryo can make channels with each
other. Most adult vertebrate cell types do, at least in culture,
except, as I already said, skeletal and most nerve fibers.

HARDIN: I'm curious about this business of the cell being
damaged on the outside. You said, as a result of increase in calcium, it closes the pores to the other cells. This would seem to
be a very noble self-sacrificing sort of impulse on the part of
that cell. Are there cases where the other cells close the pores
out of self-protection? How do you know which one it is?

LOEWENSTEIN: We know this from experiments in which a single
cell is injured under controlled conditions. We drill a microscopic hole into a cell with a micropipette, and measure electrical
communication or transfer of fluorescent molecules between that
cell and an injured neighbor, and between that neighbor and another
uninjured neighbor. Only the injured cell is then found uncoupled,
the only cell in which the calcium ion concentration rose markedly
by the injury. Fortunately, the reaction of the channel to calcium
is very fast. It is so fast that it cuts off the flux of this ion
through the channel—actually an interesting case in membrane
biology where an ion limits its own flux through the membrane. If
it were not for this fast channel closure mechanism, the interconnected cell ensemble—which often means an entire organ or large
parts of it—would soon die. You call this self-sacrifice. Well,
it needn't always be that heroic. In case of a minor injury—
something we can also manage to do experimentally—the cell membrane can eventually seal and the channels then open up again as
the injured cell gets rid of the excess calcium by its normal calcium pumps. In case of actual cell death, yes, the self-sealing
act may look a bit more heroic. Consider, for instance, what

happens towards the end of a life span of a cell in an interconnected tissue. Most of the cells in our organism, except the nerve cells, are short-lived; they last on the average not more than 10^2 days, many last less. So as the cells reach the end of their life span, their ion pumps are the first to suffer, and as the calcium pump is slowed or stands still, the calcium ion level inside the cell rises, and the channels shut. Heroics aside, were it not for this elegant self-sealing mechanism--which evolution seized early on--an interconnected tissue like our skin could not survive the death of a single cell.

FOX: I would like to mention that experimental reasons exist for believing that the kind of cell-cell interaction that we have been hearing about is very ancient indeed (Fox, Hsu, Brooke, Nakashima, and Lacey, Intl. J. Neurosci. 3 183-192, 1972).

METZ: As a reproductive biologist, after hearing Werner Loewenstein's presentation, I have to remind people about the blood-testis barrier in the mammal. The germ cells in the mammalian male are separated from the rest of the organism by a structure known as the blood-testis barrier. The ultimate level of this is the membrane of the sertoli cells isolating and insulating the germ cells from contact with other cells in the organism. And, the functional contact where metabolites that pass through are systems of tight junctions, with presumably channels such as you described and also beautifully demonstrated by Don Fawcett. The significance of the blood-testis barrier, of course, is that the differentiating male germ cells produce autoantigens, and these must be prevented from contact with the immune system. Otherwise, the immune system will respond to these autoantigens and the resulting humoral and cellular immune components will attack the germ cells, the spermatozoa, and frequently produce infertility in the male. This is beautifully specific.

LOEWENSTEIN: The germ cell system represents one of the cells, in fact, which has been shown to be connected by this type of channel and is hormonally responsive. Communications are only turned on when the proper hormone has been given to the cell system.

RUTTER: I was just wondering in that case whether you don't believe that when cells are connected like that, that the net effect is to dampen individuality almost completely and they begin to act as an organ. It seems to me that the most likely explanation for that kind of activity is that all of the regulatory molecules, which are small, at least in cell lines which we know about, and the consequences of such channels is that you have all of the cells that are so connected equilibrate, independent of the number of receptors on the outside and their response is as if the individual is really active, and acting in concert.

LOEWENSTEIN: Right, that is what I meant before by "Gleich-
schaltung" and by "acting in concert." All this applies, of course,
only to regulatory molecules that are channel-permeant, and, in the
case of receptor-activated intracellular regulation, it implies
cellular response amplification.

CELL LINE SEGREGATION AS A KEY EVENT OF

EMBRYONIC DEVELOPMENT

Alberto Monroy[1] and Elio Parisi[2]

[1]Stazione Zoologica, 80121 Napoli, Italy
[2]Instituto di Embriologia Molecolare
80072 Arco Felice, Napoli, Italy

INTRODUCTION

It is, perhaps, seemly to introduce our topic by presenting a chart from Ernst Haeckel's "Allgemeine Morphologie der Organismen" (1866), the subtitle of which was "The General Science of the Animals" (Fig. 1). Very little, and that mostly vocabulary, should be changed to compile a modern chart entitled "The Study of the Animal Individual." The eight columns at the bottom of the chart show Haeckel's conviction that the basis for a "Gesammtwissenschaft" of animals was the study of their development in its different aspects. In particular, he wrote "All morphological and physiological unions of aggregated individuals . . . are the necessary result of the simpler individuals that compose it . . . and, to be sure, in the last instance of its active constituents."

The question we propose to discuss bears on the problem of how and to what an extent the individualities of cells become integrated in the organism as a whole. Our approach will be to trace the morphological and physiological pathways of individual cells, the lineage cells, in the embryo until they reach their final destination in a specific organ.

Since all multicellular organisms originate from one cell—the fertilized egg or zygote—the problem is how the cell divisions which follow fertilization give rise to a variety of diversified cell lines rather than to a clone of identical cells. We shall present evidence that this is due to a differential segregation of cytoplasmic components into different blastomeres. These components are either already nonuniformly distributed in the unfertilized egg or are redistributed to different cytoplasmic territories

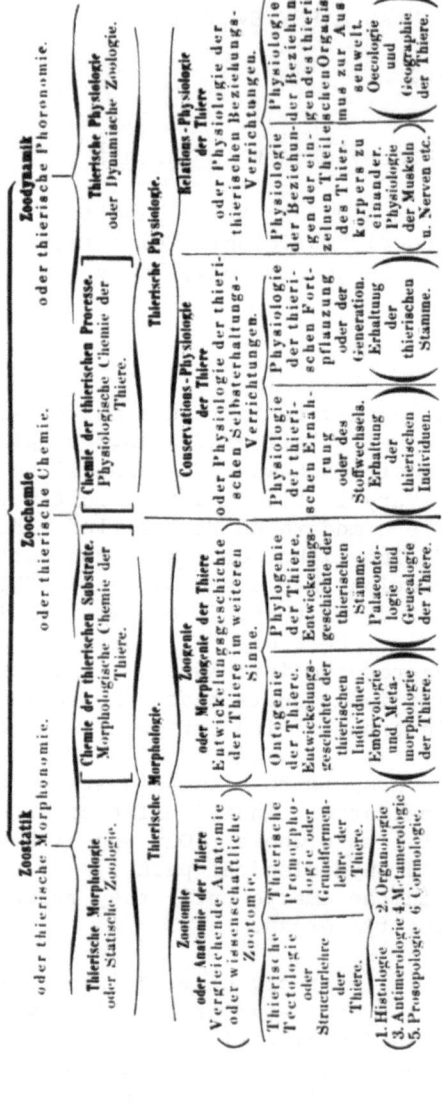

Fig. 1. The Table from Haeckel's "Allgemeine Morphologie der
Organismen" on the "General Science of the Animals."

of the zygote following fertilization. As a result, the nuclei in the individual blastomeres are exposed to different cytoplasmic environments; hence, each nucleus will be subjected to factors that differentially influence its gene expression. Differential segregation of cytoplasmic components in the blastomeres during cleavage, with the ensuing activation and repression of different sets of genes in their nuclei, results in the segregation of cell lines; i.e., of groups of cells endowed with a specific developmental program (Rossi et al., 1975).

The formation of cell boundaries during cleavage results not only in the isolation of specific cytoplasm territories, but also establishes a condition whereby adjoining blastomeres can exchange information through a communication system based on certain specializations of the cell membranes in the areas of contact. These cell-cell interactions are of fundamental importance for the coordinate differentiation of the individual cell lines. The biophysical aspects of the problem have already been discussed by W. Loewenstein.

One of the best examples of such a coordination--"orchestration," to use Loewenstein's word--is the control of the number of cell divisions in the various cell lines. Organs are made up of a definite number of cells. Studies on small animals whose organs consist of limited number of cells, which makes cell counting feasible and accurate, have shown that the range of variability of this number is rather narrow. This implies that cells must be endowed with division-counting mechanisms.

After a period of neglect, the analysis of cell lineage began to attract attention again. Indeed, new conceptual and technical advances have opened the way to the study of one of the most fascinating problems of development, namely, that of tracing the differentiative pathway of specific cells to their stem cells, and the latter to certain specific blastomeres. As an example, we should like to mention the results obtained by G. Stent and his colleagues (see, in particular, Weisblat et al., 1980), and the review by Stent and Weisblat, 1982) in the study of cell lineage in the leech. These investigators have been able to trace the pedigree of nerve and mesoderm cell clusters in each body segment to one single stem cell. They have also been able to show that the descendants of the four pairs of teloblasts, two groups of four cells that are segregated early during cleavage, are each one the founder of a distinct neuronal family.

Our discussion starts with an effort to trace the earliest appearance in unicellular organism of characters which may foreshadow the segregation of cell lines and hence the formation of multicellular organisms.

2. THE ANCESTRAL CELL LINES: GERM AND SOMA

    In the unicellular organisms, vegetative and reproductive
functions are performed by the same cell. Multicellular organisms, on the other hand, must be endowed with at least two types
of cells: the germ cells which serve the function of propagating
the species, and the somatic cells whose role is to establish the
conditions for the survival of the germ cells and hence of the
species. There are no clues, even on which to speculate, as to
how such a dichotomy arose in the course of evolution. The segregation of the two lines, the germ and the somatic, thus appears to
be an event at the very root of the evolution of multicellular
organisms. And yet, this is only one half of the problem. The
other, and in fact no less important half, concerns the processes
underlying the segregation of the female from the male cell line
within the germ line. A discussion of this topic would be outside
the scope of this presentation; the interested reader should con-
sult the book by Maynard-Smith (1978) and the papers by Parker et
al. (1972) and Parker (1978). Hence, for the time being, let us
start by taking for granted that at one point in the evolution of
multicellular organisms the two classes of gamete arose. However,
for our purpose, it is appropriate to point out some of the essential characters which differentiate the somatic from the germ cells.

    The most important difference between the two cell lines is
that the latter has acquired the ability to carry out meiosis. As
a matter of fact, in eukaryotes sexuality is inseparable from
meiosis. Again, how meiosis arose is hard to imagine (and, indeed,
it is one of the major problems of biology).

    Next comes the establishment of a condition whereby only
gametes of opposite sex can mate. Here some speculation is per-
missible because of similar conditions found in the present day
prokaryotes. In prokaryotes it is known that sexuality depends on
a sexuality factor, F, which may be present either as an episome
or as a genetic material incorporated in the bacterial genome. In
either cases, the genes carried by F confer a number of properties
to the F^+ cells, among which one of the most important is that
they code for some surface proteins (which are lacking in the F^-
cells), and which allow the mating with F^- cells while preventing
mating with F^+ cells; the latter is called surface exclusion (see
Monroy and Rosati, 1979a).

    It is tempting to speculate that surface exclusion is the
forerunner of the property of eukaryotic organisms whereby only
cells belonging to the opposite sex can mate (Monroy and Rosati,
1979a).

    Somatic cells seem to have lost this property. It is known
that histotypic and organotypic aggregates can be obtained from

cells of different genetic sex (Moscona, 1957; Moscona and Moscona, 1965) and chimarae can be produced from fused mammalian blastocysts irrespective of whether they are of the same genetic sex (Tarkowski, 1961; Mintz, 1962).

3. SEGREGATION OF THE GERM AND THE SOMATIC CELL LINES

The question of how the characters of an individual are transferred from one generation to another is central to the mechanism of evolution. This is not the place to enter into the historical background of the problem; suffice it here to refer to the review of Eddy (1975) and to the recent book of McLaren (1981).

The turning point in the study of the origin of germ cells was the discovery by Boveri (1899) of the early segregation of the germ cell line in the nematode, Ascaris megalocephala.

Fig. 2 shows the early stages of cleavage of this egg. The first cleavage results in the formation of two blastomeres, in one of which, the S (or A B) blastomere, the two large chromosomes of the zygote are fragmented into small chromosomes, part of which is then destroyed; this phenomenon has been called chromosome diminution. The A B blastomere is the stem cell of the ectoblast. In the other blastomere, P, the chromosomes retain their original shape, number and size; i.e., no chromosome diminution occurs. At the next cleavage, the P blastomere cleaves again into two blastomeres one of which undergoes chromosome diminution and becomes the stem cell of the secondary ectoblast, while the other again retains its number and size of chromosomes. The process continues through the fifth cleavage (at each cleavage one of the blastomeres undergoes chromosome diminution and becomes the stem cell of the tertiary and quaternary ectoblast) at which time both blastomeres deriving from P do not any longer undergo chromosome diminution and become the stem cells of the germ line, VG_I and VG_{II}. The segregation of the A B and P lines proceed essentially in the same way in another Nematode, Caenhorabditis elegans (Deppe et al., 1978) (Fig. 3) in which, however, no chromosome diminution occurs in the A B line. This process suggests that already at the time of the first cleavage, some cytoplasmic components, which prevent chromosomes from undergoing "diminution" is differentially segregated into the P blastomere. This has been shown by the recent observations of Strome and Wood (1982) that at each division of the P blastomeres, cytoplasmic granules specific of the germ cells are selectively segregated to the germ line precursor cells (P_1 to P_4) and are finally localized exclusively in the Z_2 and Z_3 germ line stem cells. These granules probably correspond to the electron-light areas previously observed in the P line blastomeres (Krieg et al., 1978). We suggest that this cytoplasmic component is responsible for the prevention of restriction of developmental potencies that occur in the somatic cell line.

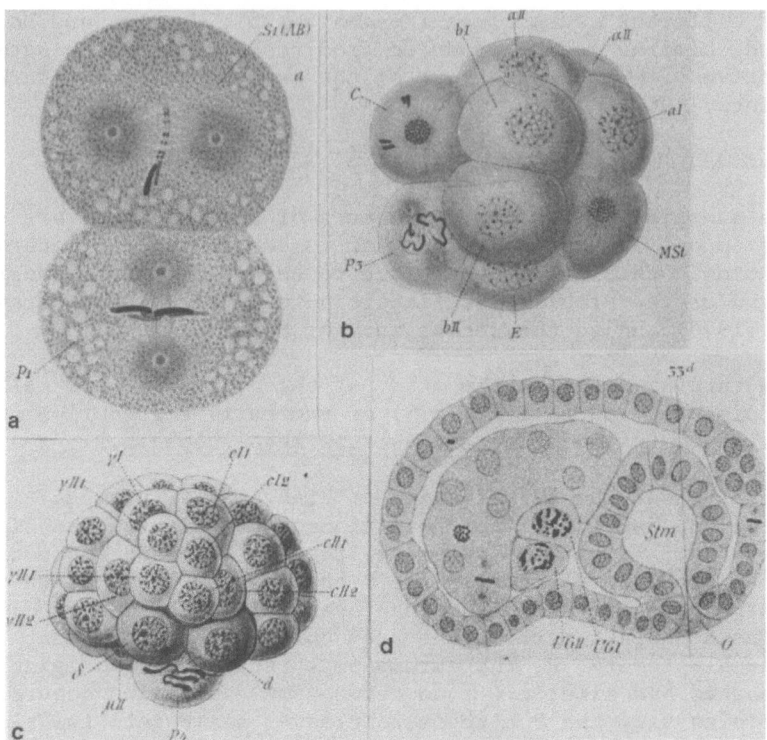

Fig. 2. Four cleavage stages of <u>Ascaris</u> showing the elimination of
part of the chromosomes in the blastomeres of the somatic (S) (AB)
line and the retention of the intact chromosome complement in the
germ line precursor blastomeres (P). (a) 2-cell stage; (b) 12-cell
stage; (c) 24-cell stage; (d) section of an embryo showing the two
stem cells of germ line, VG_I and VG_{II}. (This is a halftone copy
of the original colour illustrations of Boveri's monograph (1899).

In other eggs, such as those of some insects, a massive loss
of chromosomes in the nuclei of the somatic line occurs. In these
eggs a granular, RNA-rich material is present at the posterior pole
of the egg (see Mahowald, 1962). When the cleavage nuclei begin
their peripherad migration, those entering the pole plasm keep
their chromosome complement intact while all the others lose some
of their chromosomes (32 out of 40 in <u>Wachtiella persicaria</u>, Geyer-
Duszynska, 1959).

By centrifugation of the unfertilized egg, the RNA-containing
pole granules can be displaced into another territory of the egg.
Nuclei that enter this territory now retain their full complement

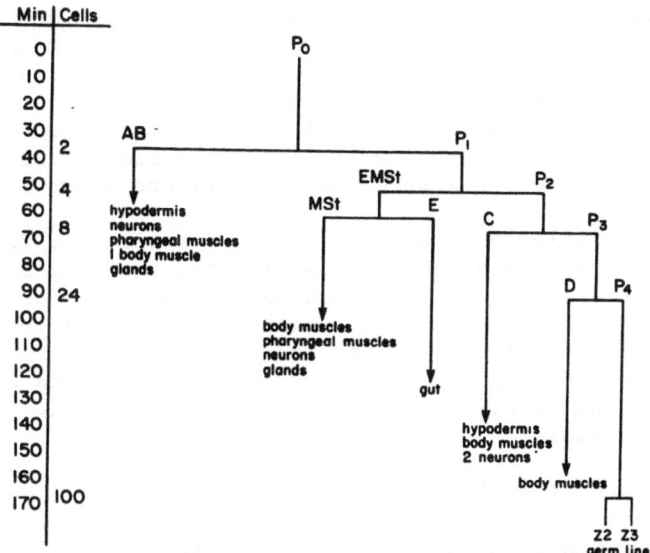

Fig. 3. Divisions of the P-cell in <u>Caenorhabditis</u> until the formation of the two germ line stem cells, Z_2 and Z_3. Source: Strome and Wood, 1982.

of chromosomes; while the pole plasm, which is now emptied of its RNA-rich granules, has lost the ability to prevent chromosome elimination. This is a strong argument that the nuclear events leading to the segregation of the germ cell line are controlled by cytoplasmic components which are in fact synthesized and segregated to a specific region of the oocyte during oogenesis.

Among the vertebrates, a similar cytoplasmic component has long been known to be present at the vegetal pole of the egg of some amphibians (reviewed by Eddy, 1975). The cells which arise from this area and hence contain such a material in their cytoplasm have been traced to the gonad. However, there is so far no evidence that the component exerts any influence on their genome. The "nuage" of the mammalian germ cells, that is considered the equivalent of the germ plasm (Eddy, 1975), has been detected in the primordial germ cells of the mouse (Spiegelman and Bennett, 1973) and of the rat (Eddy and Clark, 1975) while still in the process of their migration towards the genital ridge. However, its presence in the oocytes is not firmly established. Certainly, it has not been demonstrated in the cells of the epiblast from which the germ cells derive (see McLaren, 1981).

The segregation of the germ cells is probably the best example
of cytoplasmic control of nuclear events. It further shows that
the segregation of a cell line appears to occur in steps involving
the progressive "cleaning" of the dividing cells; which means the
differential segregation of specific cytoplasmic factors to different cells which thereby become the lineage cells. This question
will be discussed in greater detail in sections 5 and 6.

The segregation of the cells of the somatic line from those
of the germ line thus appears to involve the inactivation (whose
extreme case is the loss of chromosomes) of one part of the genome
and hence a restriction on their developmental potencies. On the
other hand, the specific character of the germ cells is that of
being prevented from undergoing such a restriction and this appears
to depend on the specific action exerted by a cytoplasmic component on their nuclei.

The molecular analysis of the process of chromosome diminution in Ascaris (Tobler et al., 1972); Moritz and Roth, 1976) has
shown that in the somatic cell line a massive elimination of
repetitive sequences occurs. It looks as though "the presence of
highly repetitive DNA components in itself is sufficient for the
genetic status of a germ line cell" (Moritz and Roth, 1976).

In this connection the results of the molecular analysis of
the development of the macronuclei in some ciliates are worth
mentioning. It is known that two nuclei are present in these protozoa; a vegetative macronucleus, and a micronucleus which is
responsible only for the meiotic and conjugation events. The
macronucleus of the ex-conjugant cells is derived from one of the
four micronuclei originating from the division of the zygotic
nucleus, the old original macronucleus having been destroyed
during the process of conjugation. By comparing the genome complexity of the micro- and of the macronucleus, the latter is found
to contain a much smaller amount of repetitive sequences than the
micronucleus. And, hence, macronuclear development is accompanied
by the elimination of a large amount of micronuclear sequences,
thus resulting in a considerable reduction in DNA sequence complexity. Once again, these observations have suggested that "large
amount of repetitious DNA . . . are not necessary for gene activity
in the macronucleus. Repetitious sequences may have a function in
the micronucleus, for example, in chromosome structure, mitosis or
meiosis" (Lauth et al., 1976; Brown, 1981, for a discussion).

Hence, this situation may be considered as a kind of ancestry
of the segregation of a somatic from a germ line.

Is the restriction of the developmental potencies a character
common to the nuclei of all somatic cells? The early experiments

of nuclear transplantation seemed to contradict such an assumption, as indeed they appeared to show that <u>any</u> somatic nucleus when injected into an enucleated oocyte was able to sustain its development. However, more recent work shows that not all somatic cells are equivalent from this point of view. For example, nuclei of epithelial cells of the intestine or of cells of adult frog tissues cultured <u>in vitro</u> have the ability to promote embryonic development of amphibian eggs through the adult (see review by Gordon, 1974).

On the other hand, in the mouse embryo the nuclei of the Inner Cell Mass (ICM) of the young blastocyst when injected into an enucleated activated egg are able to support its development while the nuclei of the trophoblast are not (Illmensee and Hoppe, 1981). This suggests that homologous genomic sites may undergo <u>irreversible</u> inactivation in some cell lines while remaining active or available for reactivation in the cells of other lines. How is this differential gene inactivation controlled and operated? This is indeed the main problem of development. We assume that it results from differential nucleo-cytoplasmic interactions which implies, as stated at the beginning of this paper, a <u>differential segregation of cytoplasmic components among the individual blastomeres during cleavage</u>.

4. THE CONTROL OF CELL DIVISION DURING CLEAVAGE

Cell division in the embryo is not a haphazard process. The more it is studied the more the pattern of cell divisions appears to be subject to tight control. In fact, it appears to be a part of the differentiation program of each cell line.

To illustrate this point, we shall discuss two examples from the recent and not so recent literature.

4a. Cell Division Counting Mechanisms

One of the most exciting results of the study of cell division and cell lineage is that the individual cell lines are each programmed for a particular and well-defined number of cell divisions.

An elegant experiment has been performed by Dan-Sohkawa and Satoh (1978) on the starfish egg. This egg is of a regulative type; until at least the 16-cell stage each isolated blastomere can develop into a normal-looking, though undersized larva. These investigators have found that the number of cells of gastrulae obtained from each of the blastomeres isolated from 2-, 4-, and 8-cell embryos respectively is roughly 1/2, 1/4, and 1/8 the number of cells of the normal gastrula.

Another well-known example is that of the micromeres of the sea urchin egg about which more will be said in section 4b. The micromeres are a group of small blastomeres that are segregated at the vegetal pole of the embryo at the 4th cleavage (Fig. 4) which are committed to giving rise to the primary mesenchyme (from which the spicules of the larva originate). It was discovered by Driesch (1898) that each sea urchin species has a fixed number of primary mesenchyme cells; and, in fact, that the micromeres are programmed for 3- and 4-cell divisions. In experiments in which the first two blastomeres were separated and allowed to develop (in the sea urchin, a normal larva can be obtained from each of the first four blastomeres) the larvae contained <u>half</u> the number of the primary mesenchyme cells.

As another example, we should like to mention some observations on the ascidian larvae. More will be said about cell lineage in the ascidians (section 5), which are in fact a classical object for this kind of study.

The data in Table 1 show that when the 64-cell stage (which is the stage when the segregation of the individual cell lines has been completed) is taken as a base line, each cell line appears to be programmed for a different number of replication cycles. This means that at this stage not only is each line committed to a well-defined differentiative program, but also the number of cell divisions they shall have to undergo <u>is a part of their differentiative program</u>. This means that the different cell lines must be endowed with very precise counting mechanisms of cell divisions. In fact, this is one of the most important morphogenetic mechanisms for the regulation of the harmonious development of the organs of the individual. Incidentally, this is likely to be one of the mechanisms which are disrupted during neoplastic transformation.

4b. The Role of Micromeres in the Control of Cell Divisions in the Sea Urchin Embryo

In the unfertilized sea urchin egg there are neither morphological nor biochemical markers of cytoplasmic compartimentalization. Nevertheless, the experiments of Hörstadius (see his book, 1973) have shown that the animal-vegetal axis of polarity is already fixed, as are also the developmental potencies of the animal and the vegetal hemispheres. It should be noted that most of these experiments have been made possible by a morphological marker which is present in some animals of the genus <u>Paracentrotus</u>, i.e., a pigment band located along the equator of the egg. This had led to the demonstration that the intestine of the larva arises from the subequatorial hemisphere of the egg (hence, the name <u>vegetal</u> half), whereas the ectoderm (i.e., the tissue which by enveloping the larva establishes the interactions with the outer world; hence the name <u>animal</u> half) originates from the supraequatorial hemisphere (Fig. 4).

Table 1. Number of cells in some of the organs of the Ascidian
          larva at the time of hatching at the 64-cell stage when
          the individual cell lines have been completely segregated
          (Data in column two from Berrill, 1935)

Number of cells in each cell line at the 64-cell stage		Number of cells in the newly hatched larva	Average number of cell divisions
Nerve cells	8	250	5
Ectoderm	26	800	5
Sense organs	2		
Endoderm	10	500	5.5
Notochord°	6	40	2.5
Mesenchyme	4	900	7.5
Muscle of the tail°	8	40	2.5

°The notochord and the tail are resorbed at the time of meta-
morphosis, just after hatching.

Now, experiments of Hörstadius (1939) have shown that when an
egg is cut into two halves along the equator and the two are fertilized, only the vegetal half develops into a normal (though undersized) larva. The animal half does not develop beyond an empty
spherical, ciliated structure named a "permanent blastula" (Fig. 5).

Fig. 4 illustrates some of the main developmental stages of
the sea urchin. The first two cleavages are meridional and give
rise to four equal blastomeres. At the fourth cleavage in the
vegetal blastomeres the cleavage plan is strongly shifted towards
the vegetal pole and this results in the formation of four small
blastomeres, micromeres, and four large blastomeres, the macromeres. At the same time, the four animal blastomeres divide
meridionally thus giving rise to a crown of eight blastomeres,
the mesomeres. The latter differentiate into the ectodermal
structures of the larva; the macromeres give rise to the gut, the
secondary mesenchyme and the coelomic sac; the primary mesenchyme
cells from which the skeleton of the larva differentiate arise
(Fig. 4) form the micromeres.

The first four cleavages are synchronous in all the blastomeres; at the fifth cleavage, the micromeres divide ahead of the
other blastomeres. The division of the first four micromeres
results in the formation of an internal and an external tier of
cells. At the following division, only the external micromeres
divide (Fig. 6) (Parisi et al., 1978). This suggests the possibility that two different cell lines originate from the division

Fig. 4. Diagram illustrating the cleavage and the early stages of
development of the sea urchin egg. A, uncleaved egg; B, 4-cell
stage; C, 8-cell stage. The B blastomeres arise from the equatorial
cleavage of the first 4 blastomeres; D, 16-cell stage. The upper,
animal blastomeres divide longitudinally, thus giving rise to 8
animal mesomeres; in the vegetal blastomeres the cleavage is hori-
zontal and the cleavage furrow is strongly shifted toward the vegetal pole resulting in 4 macromeres and 4 micromeres (black); E,
32-cell stage; F, 64-cell stage; G, early blastula with cilia; H,
blastula with flattening of the vegetal pole. In G and H the
black area indicates the position of the derivatives of the micromeres which, in I, migrate into the blastocoel, forming the primary
mesenchyme. From these cells, the skeleton of the larva arises.
K_1 and K_2, side and top view of a gastrula; note in K_1 the fully
invaginated primitive intestine, from the tip of which the cells
of the secondary mesenchyme (open circles) derive and in K_2 the
appearance of the skeletal rods attached to the cells of the primary mesenchyme; L, prism stage; M, pluteus stage from the left
side. The broken line indicates the position of the egg axis; N
pluteus from the anal side. Indication of the location of the
presumptive territories and their fate on the egg and in the cleavage stages; continuous line and dots animal territories (an_1 and
an_2); crosses and broken lines vegetal territories (veg_1 and veg_2).
aa, anal and oa, oral arms of the pluteus respectively; ar, anal
rod; vtr, ventral transverse rod, and br, body rod of the skeleton;
stom, stomodeum (primitive oral opening). From Hörstadius (1939).

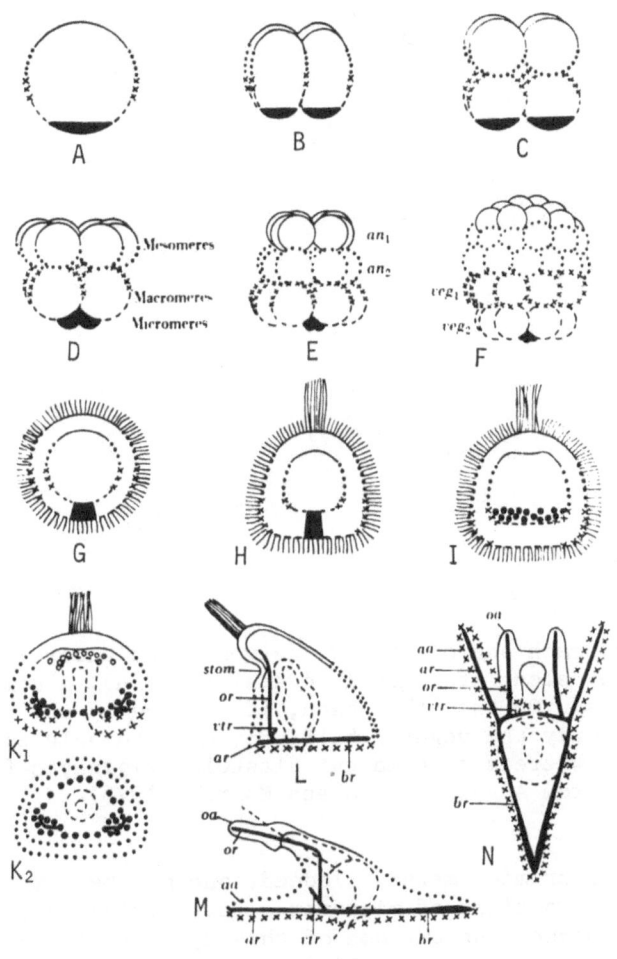

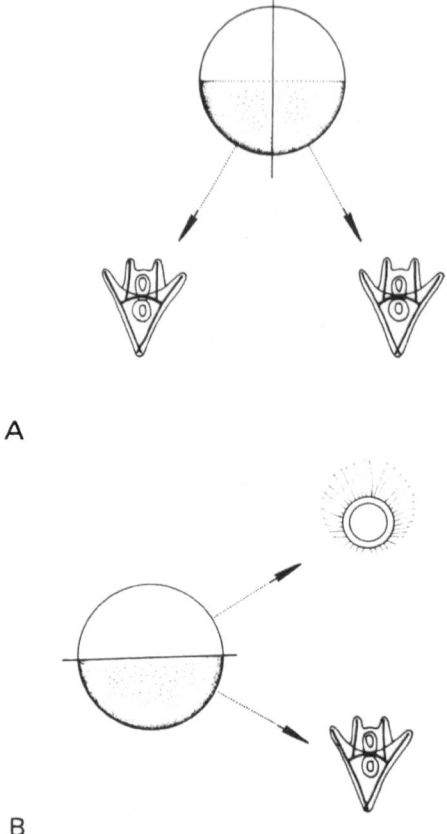

A

B

Fig. 5. From each one of the two halves őf an unfertilized sea urchin egg cut along its animal-vegetal axis (A), a complete pluteus arises. On the other hand, if the egg is cut along the equator (B), only the vegetal half proved to be able to give rise to a pluteus, whereas a permanent blastula originates from the animal half (from Augusti-Tocco and Monroy, 1975).

of the first four micromeres. Indeed, one may be tempted to specu-late that the two tiers of micromeres have different developmental potentials. Hence, the outcome of this type of cleavage is the precocious segregation of a cell line <u>committed</u> to the differentia-tion into a specific larval structure. In fact, isolated micro-meres cultured <u>in vitro</u> are able to differentiate the spicules (Okazaki, 1975). It has been suggested that during the first two cleavages a flux of surface material occurs towards the vegetal pole and eventually concentrates in the micromeres (Catalano, 1977). A similar occurrence has been observed earlier in the cleavage of the egg of the Ctenophore, <u>Beroe</u> (Spek, 1926).

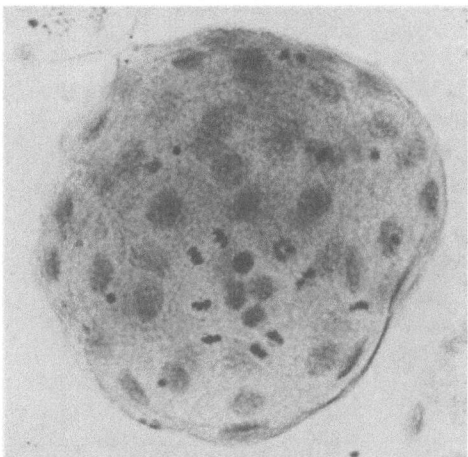

Fig. 6. A 32-cell embryo of <u>Paracentrotus lividus</u>: the nuclei of the four outer micromeres are in anaphase while the nuclei of the four inner ones are in the resting stage (from Parisi et al., 1978).

The micromeres play a most important role in the morphogenesis of the sea urchin larva. One of the most elegant evidences has been provided by the experiments of Hörstadius (1973) in which the implantation of the micromeres on top of the animal pole of a young blastula has resulted in the formation of a secondary embryo. When the micromeres were implanted into an isolated animal half of a blastula (which under normal conditions would not develop beyond the stage of a "permanent blastula" (Fig. 5), a normal blastula will develop.

Micromere segregation inaugurates a new rhythm of mitotic activity. As mentioned before, not only do the micromeres divide ahead of the other blastomeres, but the vegetal blastomeres enter mitosis prior to those in the animal hemisphere. The phase difference between dividing blastomeres becomes more pronounced at the 32- and 64-cell stage. It looks as though a propagating wave-like signal moving from the vegetal to the animal pole caused the cells to enter mitosis in a gradient fashion and, in fact, the closer they are to the micromeres the earlier they enter mitosis (Parisi et al., 1978) (Fig. 7).

The phase difference between dividing cells increases with the increase of cell number and consequently the gradient becomes steeper (Fig. 8). At the blastula stage mitoses become asynchronous and, in fact, they appear to form clusters; wave propagation of the mitotic activity can still be observed at this stage since the size of the clusters increases in colchicine-treated embryos

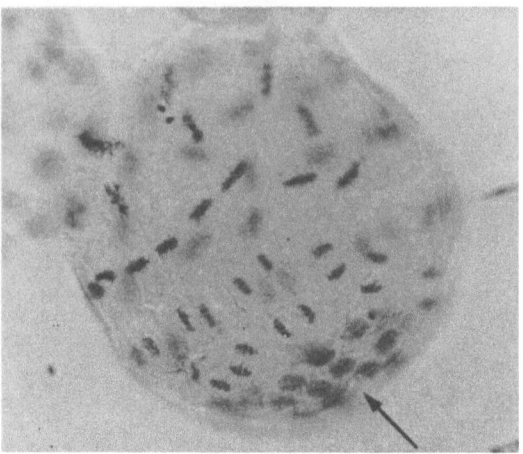

Fig. 7. A 64-cell stage embryo: the micromeres (arrow) are in interphase: vegetal blastomeres in anaphase; the animal blastomeres closer to the equator are in metaphase; those at the animal pole are in pro-metaphase (from Parisi et al., 1978).

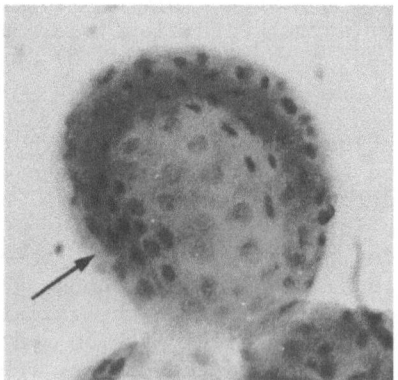

Fig. 8. In the early blastula the mitotic gradient has become steeper. Position of micromeres is indicated by arrow; the cells in the vegetal half are in interphase while those in the animal half are in late anaphase.

(Parisi et al., 1978). What is responsible for this pattern of cell division? We have suggested that cleavage is controlled by a cell division clock with the characteristics of a limit cycle oscillator; we have postulated that a signal of some kind is periodically fired within the blastomeres and triggers mitosis. Before the segregation of the micromeres the blastomeres divide at the same pace; once the micromeres have been segregated they establish their own mitotic rhythm which perturbs the synchrony of the other cells. Thereby each individual clock is now entrained by the rhythm of the micromeres which act as a pacemaker of the mitotic activity of the embryo. This behavior can be interpreted in terms of coupled limit cycle oscillators (Winfree, 1967). Coupling is an important element for the transduction of the cell division signals and is likely to depend on junctions between blastomeres. There is recent evidence that at this stage gap junctions are present between the blastomeres (Andreuccetti and Filosa, personal communication).

As cleavage proceeds, cells become smaller and the surfaces of contact between them are concomitantly reduced. This, together with the fact that junctional coupling may change in the course of development, may result in a delay of the spreading of the signal and hence the mitotic gradient becomes steeper as cleavage proceeds.

When the population becomes large enough, the delay may cause the formation of semi-autonomous groups of cells eventually developing local mitotic gradients (Parisi et al., 1978).

Further evidence of the role of the micromeres in the coordination of the mitotic activity of the early sea urchin embryo comes from experiments in which the micromere segregation was prevented by a brief exposure between the 4- and the 8-cell stage of the embryos to a detergent such as sodium lauryl sulphate. As a result of this treatment, the cleavage spindle remains in the midline of the vegetal blastomeres whereby these divide into eight cells of equal size. The embryos remain viable; however, they fail to form primary mesenchyme and to gastrulate (Tanaka, 1976). In these embryos, cell divisions remain synchronous for several division cycles. On the contrary, treatment at the 16-cell stage, i.e., after the segregation of the micromeres, does not affect the establishment of the mitotic gradient (Filosa et al., in preparation). Clearly, in the absence of the pacemaker, the mitotic rhythm of the blastomeres is not perturbed and synchrony persists.

The pacemaker activity of the micromeres can be inhibited by blocking transcriptional events that take place just at the time of their segregation (Mizuno et al., 1974). Indeed, Actinomycin D treatment of the embryos prior to the segregation of the micromeres results in the disruption of the mitotic gradient (Parisi et al., 1979).

This situation is reminescent of the effect of the polar lobe on cell division in the embryo of the mollusk, Dentalium. Removal of the polar lobe at the trefoil stage (first cleavage) abolishes the division asynchrony between blastomeres. Furthermore, as a result of the loss of the dorso-ventral difference in the division patterns, which are responsible for transition from radial to bilateral symmetry, the embryos become radialized (van Dongen and Geilenkirchen, 1975).

5. SEGREGATION OF THE MORPHOGENETIC TERRITORIES IN THE ASCIDIAN EMBRYO

Ascidians are interesting not only because they have provided one of the best examples of cell line segregation, but also because they are the lowest representatives of the phylum of Chordates, which also comprises vertebrates.

After fertilization, the egg undergoes a rapid reorganization of its cytoplasm so that before the first cleavage, it is already subdivided into different regions, each having distinct commitments (Fig. 9a). Cleavage of the egg results in the segregation of cytoplasmic components, which are present but appear to be uniformly distributed in the unfertilized egg, into different blastomeres and hence into the ensuing cell lines. The present discussion will be essentially based on the classical work of Conklin (1905) on the egg of Styela partita and the following studies of Reverberi and his colleagues (see Reverberi, 1971) on Mediterranean asicidians. In the unfertilized egg of Styela, which is laid prior to the onset of maturation, a gray yolk mass is concentrated in the center of the oocyte and is surrounded by a peripheral layer of yellow mitochondria-rich cytoplasm. In the Ciona eggs, which are currently studied in our laboratory, maturation is started in the ovary and the egg is shed with its nucleus arrested at the metaphase of the first meiotic division; in fact, the position of the meiotic spindle marks the position of the presumptive animal pole of the embryo. In this egg the mitochondria have already largely concentrated in the opposite; i.e., the presumptive, vegetal pole of the egg (R. De Santis, personal communication). Following fertilization, the yellow and clear cytoplasms stream towards the vegetal pole, thus leaving most of the yolk in the animal half, except for a small cap of clear cytoplasm which remains in the animal pole. The male pronucleus, originally located at the vegetal pole of the egg, moves towards the center of the egg following a rather well-defined path towards the presumptive dorsal region of the egg. In this movement the male pronucleus with its astral rays appears to pull the yellow and the clear cytoplasm towards the dorsal region with a layer of clear cytoplasm on top of it; thus, the yellow crescent is formed which extends well in the subequatorial region of the egg (Fig. 9a). Shortly before the

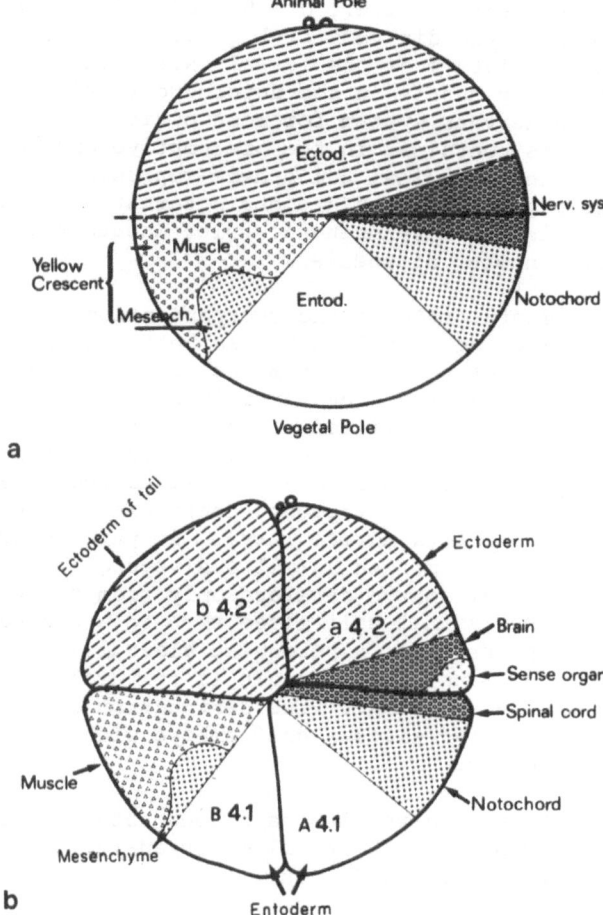

Fig. 9.   (a) Fate map in the fertilized, uncleaved ascidian egg and
(b) their segregation in the 8-cell embryo (from Monroy and Rosati,
1979b).

onset of the first cleavage, further cytoplasmic rearrangements
lead to the formation of a grey crescent opposite to the yellow
crescent, while the yolk becomes largely concentrated in the vegetal hemisphere. The animal hemisphere thus remains as a relatively
yolk-free cytoplasm. It should be noted that not in all ascidian
species can such cytoplasmic movements be visualized as neatly as
in Styela; that of the ywllow crescent is the most common identified cytoplasmic movement. The important point is that these
cytoplasmic components have been found to be markers of the

morphogenetic territories, i.e., of the individual cell lines.
Indeed, as shown in the diagrams of Fig. 9, the yellow crescent will
end up in the musculature of the larval tail. The anterior grey
crescent will end up mostly in the notochord (for its subequatorial
part) and in the nervous system for most of its supraequatorial portion. The yolk is contained in the presumptive endoderm while the
clear cytoplasm in the animal hemisphere represents the presumptive
ectoderm. The pigments, which are characteristic of the eggs of
some ascidians play no morphogenetic role; indeed they are lacking
in the eggs of a number of species.

The developmental role of the reorganization that the ascidian
egg undergoes following fertilization is better understood by comparing the development of the egg fragments obtained from the fertilized
egg with those obtained from the fertilized egg. Two halves from an
unfertilized egg, when fertilized, give rise to two essentially complete tadpoles, irrespective of the orientation of the section. The
same operation, when performed on a fertilized egg gives an entirely
different result. Only egg fragments containing the vegetal territory
are able to develop into a normal larva. In particular, as a result
of a section of an egg into an animal and vegetal half by an equatorial section, a normal embryo develops only from the vegetal half.
On the contrary, the animal half gives rise to a cluster of cells
which will never gastrulate (a permanent blastula), even when it contains the zygote nucleus. This result is comparable to that obtained
on embryos at the 2-cell stage. The destruction of one blastomere
of a 2-cell stage embryo gives a half embryo with endoderm, neuro-
ectoderm and notochord.

The most interesting results have been obtained with operations
on the 8-cell embryo in which the major organ-forming territories
have been precisely mapped (Fig. 9b).

The results of these operations show that all the pairs of blastomeres, except for the anterior animal blastomeres, differentiate
according to their presumptive fate. The anterior animal blastomeres,
when isolated, fail to differentiate into neuroblast unless they are
combined with the two anterior vegetal blastomeres (Fig. 10).

This situation is similar to that of the amphibians, where the
differentiation of the nervous system is dependent upon the chordo-
mesoderm. All the other combinations give rise either to ectodermal
vesicles or to structures containing musculature, endoderm and mesenchyme (Fig. 11).

The results of these observations and experiments thus show
that the cytoplasmic reorganization of the ascidian egg that follows fertilization turns the isotropic structure of the unfertilized egg into a mosaic of territories. The animal hemisphere is
entirely ectodermal with its anterior region containing the

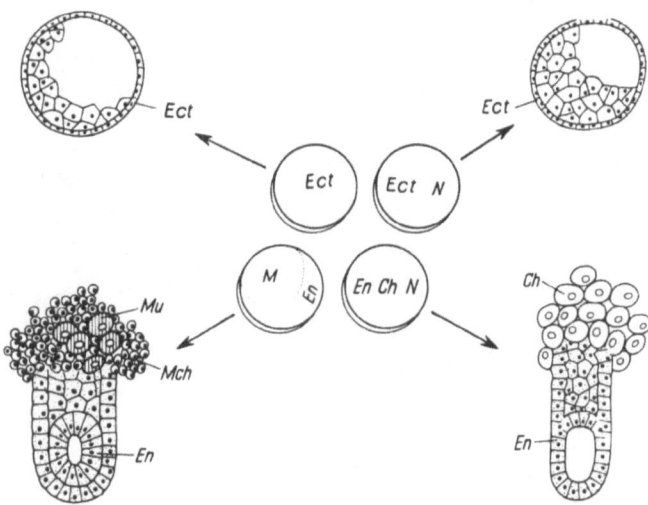

Fig. 10. Developmental fate of the four isolated pairs of blastomeres of an ascidian embryo at the 8-cell stage. Ect = ectoderm;
N = neural territory; E = endoderm; Ch = notochord; M = mesodermal
territory; Mu = musculature; Mch = mesenchyme (from Reverberi,
1971).

presumptive territory of the nervous system (which probably extends
slightly into the subequatorial hemisphere). In the vegetal hemisphere three territories can be recognized: the chordoblast,
corresponding to the grey crescent; the muscle-mesenchyme territory
corresponding to the yellow crescent (which is characterized by a
high concentration of mitochondria); and the endoderm at the vegetal pole, between the notochordal and the muscle territories. However, while the vegetal territories are already committed to differentiation, the nervous system can only differentiate under the
influence of the chordoblast.

     The segregation of the individual cell lines is completed at
the 64-cell stage (Table 1) and the segregation pathway is shown
in Fig. 11. Hence, it appears that during cleavage to the 64-cell
embryo different groups of blastomeres become progressively
enriched with specific cytoplasmic determinants (which for brevity's
sake we shall call morphogens), thereby becoming competent to
express a certain developmental program. This situation is thus
reminiscent of the segregation of the germ cell line. In this
connection we should like to mention once again the case of the
leech embryo in which the teloblasts, and only the teloblasts,
contain a cytoplasmic marker, the teloplasm. This is a clear
cytoplasm region of the egg which at the first cleavage passes

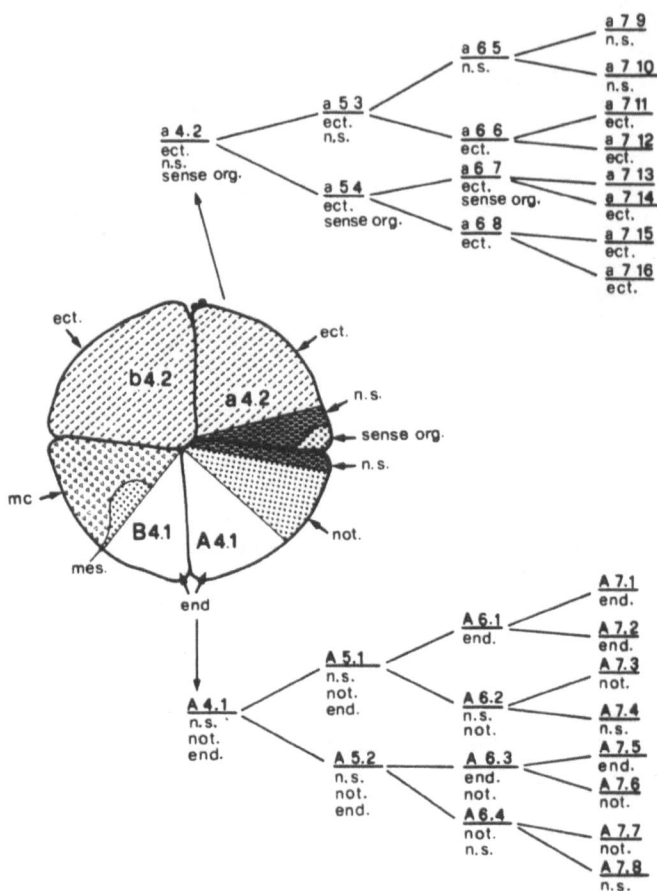

Fig. 11. Segregation of the cell lines deriving from blastomeres A 4.2 and A 4.1 from the 8-cell through the 64-cell stage. Ect = ectoderm; mc = musculature; mes = mesenchyme; end = endoderm; not = notochord; ns = nervous system; sense org. = sense organ (from Monroy et al., 1983).

almost entirely in the CD cell from which the teloblasts originate (Stent and Weisblat, 1982). Hence, here we have a case of a marker of a somatic cell line which seems to be the carrier of specific developmental information and/or signals. The nature of such "markers" is one of the most interesting and puzzling questions for the understanding of the mechanisms of commitment of cell lines.

We have recently worked out a theoretical model of the formation of the morphogenetic territories of the ascidian egg (Catalano

et al., 1981). In the model, we have interpreted the sequence
of cytoplasmic rearrangements occurring between fertilization and
cleavage as consequence of the segregation of three distinct morphogens produced by biochemical reactions, exhibiting nonlinear
kinetics. The combined action of these morphogens specifies six
presumptive territories or compartments of the embryo (Fig. 12).
Of these compartments, the one corresponding to the posterior region
of the ectoderm, which gives rise to the ectoderm of the tail, has
thus far not been recognized as an independent territory. However,
according to our prediction, it should differ from the anterior
region of the ectoderm, as to its inability to respond to the neural
induction of the notochord. Preliminary experiments in our laboratory support this hypothesis. As to the origin of the gradients,
we suggest that the first gradient, i.e. the one that moves from
the animal to the vegetal pole of the egg, may be related to the
resumption of meiosis. On the other hand, we assume that the
second gradient originates at the vegetal pole in coincidence with
the site of entrance of the fertilizing spermatozoon. Conklin
(1905) had already noted that sperm penetration in the ascidian egg
occurs in a narrow area at the vegetal pole. Current microcinemato-
graphic observations in our laboratory (R. De Santis, personal communication) on eggs deprived of their vitelline envelope, the
so-called chorion, support this conclusion. The onset of the third
gradient further modifies the constitution of the egg cytoplasm.
In particular, it results in a concentration of mitochondria in the
yellow crescent area; this may cause a variety of metabolic changes
including pH changes, calcium displacement from a bound to a free
condition and so on. As shown in Fig. 13, the combined action of
the three morphogens specifies the six compartments which were mentioned previously, specifically: (1) a region without morphogens
(the general ectoderm of the body); (2) a region containing the
second morphogen only (the presumptive territory of the ectoderm of
the tail; (3) a compartment with the first and second morphogen
(the yellow crescent): (4) an area characterized by the presence of
all three morphogens (the presumptive endodermal territory [inci-
dentally, this territory has been found to be a powerful inductor
of the nervous system (Reverberi et al., 1960)]; (5) a compartment
containing the first and third morphogens only (the presumptive
territory of the motochord); and (6) a compartment containing the
third morphogen only (the presumptive territory of the nervous system). We should like to stress the point that when speaking of the
first, second, and third gradient we do not mean to imply a sequential order of their appearance; in fact, it is likely that at
least the first and second gradients begin essentially at the same
time.

Finally, we should like to draw attention to another aspect
of ascidian development which is of interest from the phylogenetic
point of view. In all vertebrates, the roof of the archenteron is
formed by the notochord and the flanking mesoderm, whereas in the

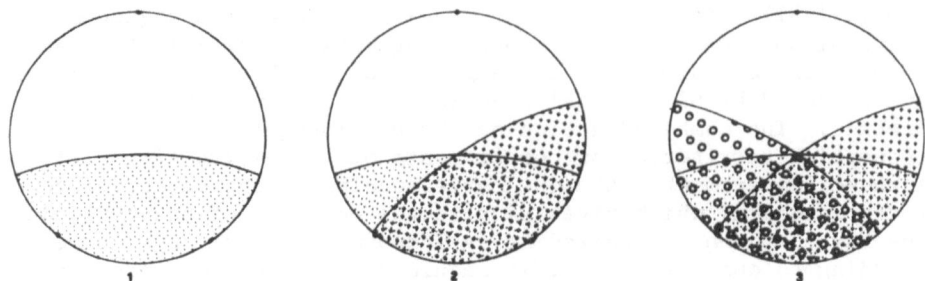

Fig. 12. Our model of the sequential formation of compartments in
the fertilized ascidian egg. 1. Threshold line of the first morphogen gradient dividing the egg into two compartments. 2. The
four compartments after the appearance of the second gradient.
The yellow crescent would correspond to the left subequatorial
section. 3. Final partitioning after the appearance of the third
gradient. The symbols denote the individual morphogens and show
those areas where the concentration of the morphogens is above
threshold level (from Catalano et al., 1979).

ascidians it is formed by the notochord and the endoderm (Fig. 13).
Comparison of the fate map of the ascidians with that of all chordates, including the cephalochordates (e.g., Amphioxus), shows
that in the ascidians the presumptive territory of the mesoderm
is restricted to the vegetal-posterior half of the egg; while in
the vertebrates it extends anteriorly underneath the territory of
the notochord, i.e., between the notochord and the endoderm.
Hence, we may assume that the situation in the ascidians is the
evolutionary forerunner of that of the vertebrates. As a matter
of fact, the idea of the ascidians' being ancestors of the vertebrates, to our knowledge, was first expressed by Charles Darwin in
The Descent of Man (1871). In his own words:

> We should thus be justified in believing that at an
> extremely remote period a group of animals existed,
> resembling in many respects the larvae of our present
> ascidians, which diverged into two great branches--
> the one retrograding in development and producing the
> present class of ascidians, the other rising to the
> crown and summit of the animal kingdom by giving
> birth to the vertebrates.

6. THE MAMMALIAN EMBRYO

Mammals pose a new problem of development which is related to
viviparity, namely implantation. Implantation is indeed an absolute requirement for the mammalian embryo to develop. Hence, one

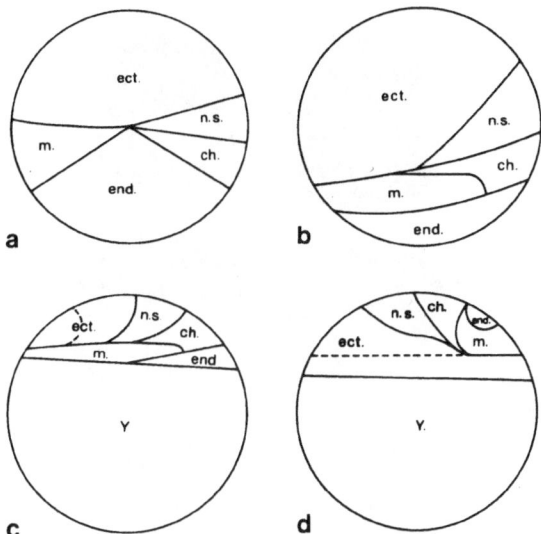

Fig. 13. Fate map of the ascidians, a; amphibians, b; the teleost fishes, c; and the birds, d. Ect = ectoderm; n.s. = nervous system; ch = notochord; m = mesoderm; end = endoderm; y = yolk (from Kuhn, 1965).

of the earliest events of mammalian development must be concerned with the preparation for implantation. As a corollary, cleavage of the mammalian egg is not immediately comparable with that of, e.g., sea urchin, ascidians, or amphibians. As a matter of fact, as will be shown presently, the first cell line to be segregated is the trophoblast, which is the structure involved in the estab- lishment of the connections between the embryo and the uterus.

The key stage in the cleavage of the mammalian egg is the 8-cell stage when the compaction begins. Compaction is marked by a number of changes that take place in the relationship between blastomeres. Indeed, at this time the surface of contact between blastomeres flattens, which results in closer apposition between them (Graham and Lehtonen, 1979). At this point, gap junctions (Magnuson et al., 1978) and electrical coupling (Lo and Gigula, 1979) between blasto- meres can be detected for the first time.

Detailed study of the cleavage pattern in the mouse egg (reviewed in Graham and Lehtonen, 1979) has shown that already at the 8-cell stage one can recognize deeply located cells with many cell-cell contacts (Fig. 14) and more superficially located cells with fewer cell-cell contacts. The situation becomes accentuated at the 12- and 16-cell stage. In the 16-cell morula there are 8 to 10 outside and 6 to 8 inside cells (Johnson and Ziomek, 1981a).

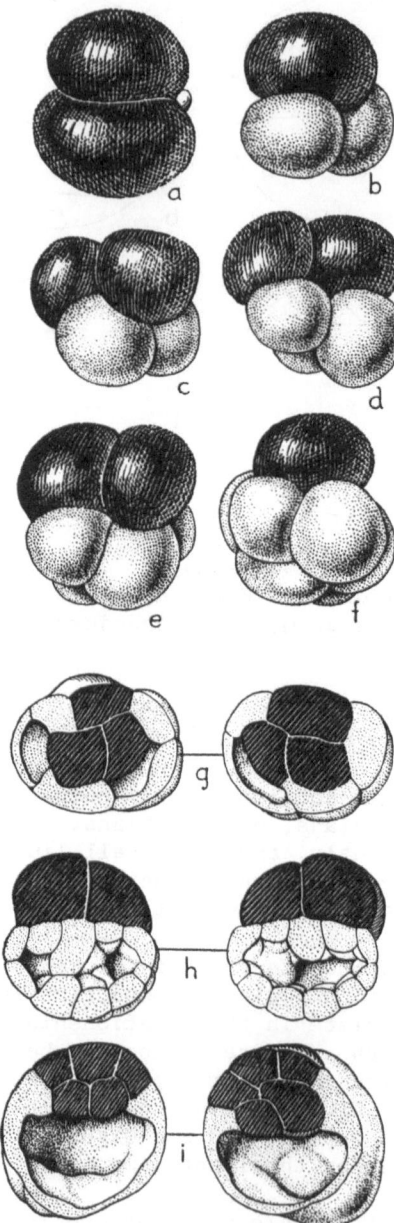

Fig. 14. Wax plate model of the cleavage of the pig egg. a to f,
total view; g to i, median sections. The more slowly dividing
cells (black) will become the Inner Cell Mass from which the embryo
will arise and the faster dividing cells (gray) will give rise to
the trophoblast (from Heuser and Streeter, 1929).

Hence, the segregation of two distinct cell populations: one taking a deep position to become the Inner Cell Mass (ICM) and the other taking an outer position to become the trophoblast, takes place very early during cleavage (Fig. 14). It has also been suggested that the orientation of the cleavage planes is the critical factor in this process and this in turn is controlled by cell contacts (Graham and Lehtonen, 1979).

At the 8-cell stage, also, the surface of the blastomeres becomes polarized; up to the early 8-cell stage the surface of the blastomeres is uniformly stained by fluorescently labeled Con A. From the mid-8-cell stage, in the majority of the cells the staining is restricted to the outward facing pole (Johnson et al., 1981; Johnson and Ziomek, 1981a). Thus, at this time a radial mosaicism is imparted to the cells (Johnson, 1981), whereby in the 16-cell morula the cells containing the polarized outer region of the 8-cell stage are segregated to the outside and become the founders of the trophoblast cell line. Polarization depends on cell contact; indeed, isolated 8-cell stage blastomeres do not polarize (Ziomek and Johnson, 1980). Cell contact is hence the polarization-inducing signal in the 8-cell morula. It is also interesting that the ability to induce polarization first appears in the blastomeres of the 2-cell embryo but the ability to respond to the polarizing stimulus appears only in the blastomeres of the 8-cell embryo (Johnson and Ziomek, 1981b). Polarization thus appears to involve a change in the properties of the membrane of the blastomeres. However, although many factors have been proposed, it is difficult to pinpoint the one(s) that may be considered as critical (Johnson, 1981).

The surface changes of the blastomeres responsible for compaction are under genetic control. There is strong evidence that genes in the T/t locus of the mouse have a critical role in development and, in fact, that mutations in this locus affect the organization of the cell surface thus upsetting development at well-defined stages (see Bennett et al., 1972; Artzt and Bennett, 1975). In the context of the present discussion, it is thus of special interest that in mouse embryos homozygous for the t^{12} allele, development is arrested at the morula stage and, in fact, at the time of compaction (Smith, 1956).

It should also be added that there is essentially no random mixing of blastomeres during cleavage; i.e., the descendants of each one of the first 5 blastomeres have a fixed position within the embryo (Kelly, 1979).

These results thus suggest that the 8-cell stage is a turning point in the development of the mouse embryo. The experimental results of mixing of blastomeres from the same or from different embryos also show that at least until that stage the blastomeres

are totipotent (Tarkowski and Wrobleska, 1967; Kelly, 1977). The
trophoblast is thus the first cell line to be segregated in the
mammalian embryo and its cells become rapidly specialized and committed. This is shown by experiments in which the developmental
potential of the trophoblast and of the ICM cells has been tested
by their transplantation into the blastocoelic cavity of a host
embryo. While the trophoblast cells appear to be unable to give
rise to anything but trophoblast, the ICM cells prove to be able to
participate in the formation of both embryonic and extraembryonic
structures (Gardner, 1974, quoted by Eddy et al., 1981). This finding has been further supported by experiments in which ICM isolated
immunosurgically from half expanded to fully expanded blastocysts
and cultured in vitro differentiate not only embryonic but also
extra-embryonic structures (Hogan and Tilly, 1978).

    Nuclear transplantation experiments (Illmensee and Hoppe,
1981) show a restriction of the developmental potencies of the
nuclei of trophoblast cells. Indeed, nuclei of trophoblast cells
are unable to support development of enucleated activated mouse
eggs while nuclei of the ICM cells do. There are also distinct
differences in the pattern of polypeptides synthesized by the ICM
and by the trophoblast (Van Blerkom et al., 1976). This, together
with the polyploidization the trophoblast cells undergo (Barlow
and Sherman, 1972), shows that the differentiation of the trophoblast begins very soon after its segregation.

    Hence, even at such late stage as the blastocyst, the cells
of the ICM still retain a wide range of developmental potencies.
This could have been anticipated from the early work of Tarkowski
(1961) and Mintz (1962) showing that mixtures of blastomeres from
32-cell blastocysts were able to give rise to normal mice. The
next cell line to be segregated is the endoderm, and here again
segregation is accompanied by a restriction of the developmental
potencies of these cells (Gardner and Papaioannou, 1975). Thus,
the mammalian embryo proves to be an excellent model for the study
of some of the events underlying the segregation of cell lines.
In particular, the study of the mammalian embryo has provided clear
evidence that the segregation of a cell line is accompanied by a
restriction of the developmental potencies of the constituent cells
and the silencing of sets of genes.

7. SOME CONCLUDING REMARKS

    After fertilization, the egg resumes the ability to replicate
its DNA and to divide. In contrast to the division of cells in
culture, the division of the egg gives rise to cells which from
very early during development become progressively restricted in
their developmental capabilities, and hence are able to express a
limited and specific developmental program. This is the result

of the inactivation of sets of genes which progresses stepwise and differentially in the different blastomeres and in the cell lines deriving from them. What are the factors controlling such a differential gene inactivation? This question is indeed central to understanding differentiation as a whole. We assume that the control is exerted by the interaction of the genome with specific cytoplasmic factors. This assumption, though very likely, still needs to be proved and the nature of the factors is still unknown. On the other hand, controls are not only intrinsic and limited to each individual cell; in other words, the differentiation of each individual lineage cell is not simply the result of the unfolding of an individual and independent program. The program comprises a complex series of cell-cell interactions and these not only with the cells of the same line but also with those of other cell lines. One of the most important aspects of the segregation of the cell, lines is that once committed to a certain developmental pathway, i.e., to a cell line, each cell is also committed to a defined number of cell divisions. The program for cell divisions is hence an essential part of the differentiative program of each cell line.

A cell line of special interest is the germ line whose segregation must have been one of the earliest events in the evolutionary history of multicellular organisms. At a variance with the somatic cells, the cells of the germ line retain the unrestricted developmental potencies of the zygote. We suggest that this process is not fundamentally different from that of the segregation of the somatic line; i.e., it is also the result of specific nucleocytoplasmic interactions. The only, indeed very important, differences are that in this case the active factors (a) prevent the restriction of the developmental potencies of the genome, and (b) confer to the gonocytes the ability to carry out meiosis. Meiosis is indeed the marker of the germ line (Monroy et al., 1983). In the case of the cells of the germ line, putative cytoplasmic factors have been identified which prevent the nuclei undergoing the changes they undergo in the somatic cell line. What these factors are and how they influence the genome of the germ line is not yet understood.

REFERENCES

Artzt, K., and Bennett, D, 1975, Analogies between embryonic (T/t) antigens and adult major histocompatibility (H-2) antigens. Nature 256, 545.

Augusti-Tocco, G., and Monroy, A., 1975, Gene-controlled regulatory processes in the development of marine organisms. Biochem. Biophys. Persp. Marine Biol. 2, 291-341.

Barlow, P. W., and Sherman, M. I., 1972, The biochemistry of differentiation of mouse trophoblast: Studies on polyploidy. J. Embryol. Exp. Morphol. 27, 447-465.

Bennett, D., Boyse, E. A., and Old, L. J., 1972, Cell surface
    immunogenetics in the study of morphogenesis. In: Cell
    Interactions, L. Silverstrini, ed. (North-Holland), 247-263.
Berrill, N. J., 1935, Studies in Tunicate development. 3. Differ-
    ential retardation and acceleration. Phil. Trans. Royal
    Soc. London B 225, 255-326.
Boveri, T., 1899, Die Entwicklung von Ascaris megalocephala mit
    besonderer Rücksicht auf die Kernvehältnisse. Festsch.
    F. C. von Kupfer (Jena), 383-430.
Boveri, T., 1910, Die Potenzen der Ascaris Blastomeren þei abgeän-
    derter Furchung. Zugleich ein Beitrag zur Frage qualitativ-
    ungleicher Chromosomen Teilung. Festsch. sechzigt Geb. R.
    Hertwig 3, 131-214.
Brown, D. D., 1981, Gene expression in eukaryotes. Science 211,
    667-674.
Catalano, G., 1977, "Cleavage fields": Hypothesis on early embry-
    onic development. Cell Differ. 6, 111-118.
Catalano, G., Eilbeck, C., Monroy, A., and Parisi, E., 1981, A
    model for early segregation of territories in the ascidian
    egg. In: Cell Lineage, Stem Cells, and Cell Determination,
    N. Le Douarin, ed. (Elsevier/North Holland), 15-28.
Conklin, E. G., 1905, The organization and cell lineage of the
    ascidian egg. J. Acad. Natur. Sci. 13, 1-119.
Dan-Sohkawa, M., and Satoh, N., 1978, Studies on dwarf larvae
    developed from isolated blastomeres of the starfish, Asterina
    pectinifera. J. Embryol. Exper. Morphol. 46, 171-185.
Darwin, C., 1871, The Descent of Man and Selection in Relation to
    Sex. J. Murray, London.
Deppe, U., Schierenberg, E., Cole, Th., Krieg, C., Schmitt, D.,
    Yoder, B., and von Ehrenstein, G., 1978, Cell lineages of the
    embryo of the nematode Caenorhabditis elegans. Proc. Natl.
    Acad. Sci. USA 75, 376-380.
Driesch, H., 1898, Von der Beendigung morphogener Elementar-
    processe. Arch. Entw. Mech. d. Organ 6, 198-227.
Eddy, E. M., 1975, Germ plasm and differentiation of the germ cell
    line. Int. Rev. Cytol. 43, 229-280.
Eddy, E. M., and Clark, J. M., 1975, Electron microscopic study of
    migrating primordial germ cells in the rat. In: Electron
    Microscopic Concepts of Secretion, M. Hess, Ed. Wiley & Sons,
    New York, 151-167.
Eddy, E. M., Clark, J. M., Gong, D., and Fenderson, B. A., 1981,
    Origin and migration of primordial germ cells in mammals.
    Gamete Res. 4, 333-362.
Gardner, R. L., and Papaioannou, V. E., 1975, Differentiation in
    the trophectoderm and inner cell mass. In: The Early Devel-
    opment of Mammals, M. Balls and A. E. Wild, eds., Cambridge,
    University Press, 107-132
Geyer-Duszynska, I., 1959, Experimental research on chromosome
    elimination in Cecidomydae (Diptera). J. Exp. Zool. 141,
    391-447.

Graham, C. F., and Deussen, Z. A., 1978, Features of cell lineage in preimplantation mouse development. J. Embryol. Exp. Morphol. 48, 53-72.

Graham, C. F., and Lehtonen, E., 1979, Formation and consequence of cell patterns in preimplantation mouse development. J. Embryol. Exp. Morphol. 49, 277-294.

Gordon, J. B., 1974, The Control of Gene Expression in Animal Development. Harvard University Press, Cambridge.

Haeckel, E., 1866, Allgemeine Morphologic der Organismen. G. Reimer Vetlag, Berlin.

Heuser, C. H., and Streeter, G. L., 1929, Early stages in the development of pig embryos, from the period of initial cleavage to the time of the appearance of the limb buds. Publ. Carnegie Inst. Contr. to Embryol. 109, 1-30.

Hogan, B., and Tilly, R., 1978, In vitro development of inner cell masses isolated immunosurgically from mouse blastocysts. 1. Inner cell masses from 3-5 day p.c. blastocysts incubated for 24 h before immunosurgery. J. Embryol. Exp. Morphol. 45, 93-105.

Hörstadius, S., 1939, The mechanisms of sea urchin development studied by operative methods. Biol. Rev. 14, 132-179.

Hörstadius, S., 1973, Experimental Embryology of Echinoderms. Clarendon Press, Oxford.

Illmensee, K., and Hoppe, P. C., 1981, Nuclear transplantation in Mus musculus: Developmental potential of nuclei from preimplantation embryos. Cell 24, 9-18.

Johnson, M. H., 1981, The molecular and cellular basis of preimplantation mouse development. Biol. Rev. Cambridge Phil. Soc. 56 463-498.

Johnson, M. H., Pratt, H. P. M., and Handyside, A. H., 1981, The generation and recognition of positional information in the preimplantation mouse embryo. in: Cellular and Molecular Aspects of Implantation, S. R. Glaser and D. W. Bullock, eds. Plenum Press, New York, 55-74.

Johnson, M. H., and Ziomek, C. A., 1981a, The foundation of two distinct cell lineages within the mouse morula. Cell 24, 71-80.

Johnson, M. H., and Ziomek, C. A., 1981b, Induction of polarity in mouse 8-cell blastomeres: Specificity, geometry and stability. J. Cell Biol. 91, 303-308.

Kelly, S. J., 1977, Studies of the developmental potential of 4- and 8-cell stage mouse blastomeres. J. Exptl. Zool. 200, 365-376.

Kelly, S. J., 1979, Investigations into the degree of cell mixing that occurs between the 8-cell stage and the blastocyst stage of mouse development. J. Exptl. Zool. 207, 121-130.

Krieg, C., Cole, T., Dieppe, U., Schierenberg, E., Schmitt, D., Yoder, B., and von Ehrenstein, G., 1978, The cellular anatomy of embryos of the nematode, Caenorhabditis elegans: Analysis and reconstruction and serial section electron micrographs. Dev. Biol. 65, 193-215.

Kuhn, A., 1965, Vorles. über Entwicklungsphysiologie. Springer, 1976.

Lauth, M. R., Spear, B. B., Heuman, J., and Prescott, D. M., 1976, DNA of ciliated protozoa: DNA sequences diminution during macromolecular development of Oxytricha. Cell 7, 67–74.

Lo, C. W., and Gilula, N. B., 1979, Gap junctional communication in the post-implantation mouse embryo. Cell 18, 411–422.

Magnuson, T., Jacobson, J. B., and Stackpole, C. W., 1978, Relationship between intercellular permeability and junction organization in the preimplantation mouse embryo. Dev. Biol. 67, 214–224.

Mahowald, A. P., 1962, Fine structure of pole cells and polar granules in Drosophila melanogaster. J. Exptl. Zool. 151, 201–208.

Maynard-Smith, J., 1978, The Evolution of Sex. Cambridge University Press.

McLaren, A., 1981, Germ cells and soma: A new look at an old problem. Yale University Press, New Haven.

Mintz, B., 1962, Experimental recombination of cells in the developing mouse: Normal and lethal mutant genotypes. Amer. Zool. 2, 541.

Mizuno, S., Lee, Y. R., Whiteley, A. H., and Whiteley, H. R., 1974, Cellular distribution of RNA populations in 16-cell stage embryos of the sand dollar, Dendraster excentricus. Dev. Biol. 37, 18–27.

Monroy, A., Parisi, E., and Rosati, F., 1983, On the segregation of the germ and somatic cell lines in the embryo. Differ. 23, 179–183.

Monroy, A., and Rosati, F., 1979a, The evolution of the cell-cell recognition system. Nature 278, 165–166.

Monroy, A., and Rosati, F., 1979b, Cell surface differentiations during early embryonic development. Curr. Topics Dev. Biol. 13, 45–69.

Moritz, K. B., and Roth, G. E., 1976, Complexity of germline and somatic DNA in Ascaris. Nature 259, 55–57.

Moscona, A. A., 1957, The development in vitro of chimeric aggregates of dissociated embryonic chick and mouse cells. Proc. Natl. Acad. Sci. USA 43, 184–194.

Moscona, M. H., and Moscona, A. A., 1965, Control of differentiation in aggregates of embryonic skin cells: Suppression of feather morphogenesis by cells of other tissues. Dev. Biol. 11, 402–433.

Okazaki, K., 1975, Spicule formation by isolated micromeres of the sea urchin embryo. Amer. Zool. 15, 567–581.

Parisi, E., Filosa, S., De Petrocellis, B., and Monroy, A., 1979, The pattern of cell division in early development of the sea urchin, Paracentrotus lividus. Dev. Biol. 65, 43–49.

Parker, G. A., 1978, Selection on non-random fusion of gametes during the evolution of anisogamy. J. Theor. Biol. 73, 1–28.

Parker, G. A., Baker, R. R., and Smith, V. G. F., 1972, The origin
and evolution of gamete dimorphism and the male-female phe-
nomenon. J. Theor. Biol. 36, 529-553.
Reverberi, G., 1971, Ascidians, in: Experimental Embryology of
Freshwater Invertebrates, G. Reverberi, ed., pp. 507-535.
Elsevier/North Holland.
Reverberi, G., Ortolani, G., and Farinella-Ferruzza, N., 1960,
The causal formation of the brain in the Ascidian larva.
Acta Embr. Morphol. Exp. 3, 296-336.
Rossi, M., Augusti-Tocco, G., and Monroy, A., 1975, Differential
gene activity and segregation of cell lines: An attempt at a
molecular interpretation of the primary events of embryonic
development. Quart. Rev. Biophys. 8, 43-119.
Smith, L. J., 1956, A morphological and histochemical investigation
of a preimplantation lethal (t^{12}) in the house mouse. J. Exp.
Zool. 132, 51-83.
Spek, J., 1926, Ueber gesetzmässige Substanzverteilung bei der
Furchung des Ctenophoreneies und ihre Beziehungen zu den
Determinationsproblemen. Arch. Entw. Mech. Org. 107, 54-73.
Spiegelman, M., and Bennett, D., 1973, A light- and electron-micro-
scopic study of primordial germ cells in the early mouse
embryo. J. Embryol. Exp. Morphol. 30, 97-118.
Stent, G. S., and Weisblat, D. A., 1982, The development of a simple
nervous system. Scient. Amer. 246(1), 100-110.
Strome, S., and Wood, W. B., 1982, Immunofluorescence visualization
of germ line specific cytoplasmic granules in embryos, larvae
and adults of Caenorhabditis elegans. Proc. Natl. Acad. Sci.
USA 79, 1558-1562.
Tanaka, Y., 1976, Effects of the surfactants on the cleavage and
further development of the sea urchin embryos. 1. The inhibi-
tion of micromere formation at the fourth cleavage. Dev.
Growth and Diff. 18, 113-122.
Tarkowski, K. A., 1961, Mouse chimeras developed from fused eggs.
Nature 190, 857-860.
Tarkowski, K. A., and Wrobleska, J., 1967, Development of blasto-
meres of mouse eggs isolated at the 4- and 8-cell stage. J.
Embryol. Exp. Morphol. 18, 155-180.
Tobler, H., Smith, K. D., and Ursprung, H., 1972, Molecular aspects
of chromatin elimination in Ascaris lumbricoides. Dev. Biol.
27, 190-203.
Van Blerkom, J., Barton, S. C., and Johnson, M. H., 1976, Molecular
differentiation in the preimplantation mouse embryo. Nature
259, 319-321.
Van Dongen, C. A. M., and Geilenkirchen, W. L. M., 1975, The development of Dentalium with special reference to the significance
of the polar lobe. IV. Division chronology and development
of the cell pattern in Dentalium dentale after removal of the
polar lobe at first cleavage. Proc. Kon. Nederl. Akad.
Wetensch. Amsterdam, Ser. C 78, 358-375.

Weisblat, D. A., Harper, G., Stent, G. S., and Sawyer, R. T., 1980,
    Embryonic cell lineages in the nervous system of the Glos-
    siphoniid leech Helobdella triserialis. Dev. Biol. 76, 58-78.
Winfree, A. T., 1967, Biological rhythms and the behaviour of popu-
    lations of coupled oscillators. J. Theor. Biol. 16, 15-42.
Ziomek, C. A., and Johnson, M. H., 1980, Cell surface interaction
    induces polarization of mouse 8-cell blastomeres at compaction.
    Cell 21, 935-942.

DISCUSSION

BOUCHARD: I am not familiar with this work at all. The question that arises in my mind is the following: normally with a human being you can continually subdivide a fertilized egg and get multiple births. You can get twins, triplets, quadruplets, etc. I have no reason to believe that these are not complete individuals. You've implied that if you divide these particular organisms, you get half the number of cells. Now, at what stage in evolution does the process shift to what we observe with humans?

MONROY: Well, that's difficult to answer. I couldn't say which stage in evolution. In the starfish, Dan and Satoh have shown quite clearly that up to the 16-cell stage, you get complete embryos from single isolated cells; in other words, from individual cells that constitute 1/2, 1/8, or even 1/16 of what was the original ovum.

METZ: There is an old concept called the nuclear-cytoplasmic ratio, which regulates cell division. Do you suppose any of that operates in these systems?

MONROY: I don't know. This is again a debatable question.

METZ: You could apply this to shutting off further micromere division, for example, after a certain number of divisions.

MONROY: Maybe, but when the size of the micromeres is compared to the size of the primary mesenchyme cells, the difference in size is minimal.

LOEWENSTEIN: What interested me very much, Alberto, was your statement that cells must be counting their number. I have one question and then I will make a comment on this. In those cases where you subdivide the system, does the rate of division actually increase?

MONROY: No.

LOEWENSTEIN: The rate is constant?

MONROY: Yes, at least in the animals I know.

LOEWENSTEIN: Yes. I think here I can see again the power of an interconnected cell system.

MONROY: I add some information that I wanted to give you yesterday. When micromeres arise, micromeres are the first cells in the embryo which show gap junctions. We haven't published that, but it's in process.

LOEWENSTEIN: Consider a system of interconnected cells. Assume that all of the blastomeres are connected via cell-to-cell channels. They have thus a common boundary. Each cell has given up individuality; the cell ensemble is the individuum. The ensemble is a system of finite volume, hence--and this is its beauty--it can get cues of its size on the basis of simple chemical concentrations.

MONROY: Diffusions. And that's what you and we have suggested independently.

LOEWENSTEIN: Assume now that each cell can put out a signal molecule--this is the model most parsimonious in degrees of differentiation. The signals dilute as they diffuse through the system in proportion to the system's aggregate cell column. Now, the system can in principle count its number and self-regulate its growth given certain conditions: (1) asynchrony in the signal generation; (2) brief durations of signal generation, durations of similar order or shorter than the signal diffusion times through the signal. This is the "dilution model" of cellular growth control I put forth some 13 years ago [Loewenstein, Devel. Biol. 19 (Sum. 2), 151-183, 1968; for a more recent version, see Loewenstein, Biochim. Biophys. Acta. Cancer Rev. 560, 1-65, 1979].

MONROY: Excuse me a minute. We consider the micromeres as pacemakers. That's a word we have used.

LOEWENSTEIN: Then you have all the elements of a dilution system. Let's say, for instance, one cell here is the one which is the pacemaker, but it doesn't have to be a pacemaker. Any cell can do it; it just has to do it synchronously, put out the signal. So in steady state that signal will dilute out and we'll say we have, on the average, two such molecules in such an ensemble. Then if the ensemble grows, as by cell division, the signal will be diluted out. So now, the only thing we need is some kind of comparison standard which will stay below that concentration level, will no longer divide, and you have all elements for controlling those and counting.

MONROY: And I may add, you probably remember that, until a certain stage, the total volume of the embryo doesn't grow. There is no increase in total volume. What increases is the number of cells. The greater the number of cells the steeper the gradient is. That's what you would assume.

METZ: What happens to the rate of cleavage when the egg is permitted to develop in calcium-free seawater where the blastomeres separate? Do they divide at the same rate as when they're in normal seawater?

MONROY: That I cannot answer. I think it has been done, but I really don't know.

METZ: That would be a nice gesture on Werner's concept.

LOEWENSTEIN: What was that?

METZ: You place the sea-urchin egg in calcium-free seawater and it continues to divide but the cells separate.

LOEWENSTEIN: Yes. You actually don't have to separate them; it would be sufficient to close the channel off, but your experiment is more rotund. One would expect that the system continues dividing until the original volume; the total aggregate cell volume has been restored. Otherwise, my model would be wrong.

MONROY: Although in the experiment that has been done by Marina Dan, when she separates the individual eight blastomeres of the starfish embryo, she gets eight larvae, each one of which is made up of 1/8 the number of cells the normal larva should have. That's why I insist on counting. But the size of the cells is identical to the one in the normal embryo.

THE INDIVIDUALITY OF ANTIBODIES

J. Donald Capra

Southwestern Medical School
The University of Texas Health Science Center at Dallas
5323 Harry Hines Boulevard
Dallas, Texas 75235

INTRODUCTION

The synthesis of an antibody in response to the administration of an antigen is an adaptive phenomenon capable of exacting, yet wide ranging, specificity. For over fifty years it has been known that antibodies are globular proteins of varying size, the majority of which are associated with the gamma fraction of the serum proteins. The term immunoglobulin is of more recent origin and serves to emphasize the basic unit of structure and function underlying the varied forms of reactivity and seemingly endless variety of antibodies.

Immunoglobulins derive from lymphoid cells generally from those terminally differentiated lymphoid cells termed plasma cells. These cells, once sensitized by antigen, become "committed" and are referred to as immunocytes. Lymphocytes are divided into two types: one concerned with cell-mediated immunity (T-lymphocytes) and the other with humoral immunity (B-lymphocytes). In general, for antibody production to take place, both B- and T-lymphocytes are required. Early lymphocytes of the B-lymphocyte lineage contain immunoglobulin on their cell surface. These immunocytes are not normally secreting cells, but undergo differentiation to plasma cells--cells specialized for secretion of antibodies. Antibodies are largely made in the concentrated lymphoid organs of the body, particularly those that contain large concentrations of B-lymphocytes, the spleen and lymph nodes. The secretory immune system which is largely concentrated beneath endothelial surfaces such as along the respiratory tract and gastrointestinal tract represents almost a separate immune system in that both the mode of antigen presentation, the class and composition of the antibody

127

molecule itself (secretory IgA) as well as its local transport into
the gut and respiratory "tubes" are unique. However, it still must
cope with an identical problem, that is, the antigen binding function must still be solved by the genetic apparatus of the individual.

Early attempts to Explain Antibody Diversity

To account for the extremely diverse range of antigens against
which antibodies can be elicited, the Instructive or Template Theory
of antibody formation, put forward in 1930, stated that an antibody
molecule is synthesized in the presence of the antigen which acts
as a template. In the 1940s this theory was extended to the notion
that all the antibody molecules of a given organism have an identical primary structure but differ from each other in the conformation of their polypeptide chains. This view was initially supported
by the finding that various antibodies have an almost identical
amino acid composition. In the early 1950s, however, partition
chromatography indicated that gamma-globulin was a complex mixture
of similar components and the N terminus of antibodies was hetero-
geneous.

These and other considerations led to the Subcellular Selection
Theory which proposed that the specificity of each antibody molecule
was determined by a unique sequence of amino acids, the diversity
arising from a high degree of spontaneous mutation.

The demonstration in the late 1950s that immunoglobulins have
two subunits called heavy and light chains (both of which are
needed for antigen binding), and similar studies on these subunits
led to the elucidation of the peptide chain structure of immunoglobulins in the early 1960s. At the same time, it was shown that
the Bence-Jones proteins, products of a cancer of the lymphoid
cells, are homogeneous proteins representative of the light chains
of immunoglobulin.   This observation provoked intensive amino acid
sequence studies on Bence-Jones proteins and culminated in a comparison of the sequence of two such proteins in 1965. The most
striking feature of these first complete amino acid sequences of
immunoglobulin light chains was that the entire carboxy terminal
half was identical in both chains (constant region), the considerable variance between the chains being confined to the amino terminal half (variable region). These observations led to one of
the most radical of ideas ever to emerge in modern biology: two
noncontiguous structural genes encode each chain.

As more amino acid sequence data accumulated, immunologists
were faced with a striking fact; every immunoglobulin studied had
a unique amino acid sequence. The origin of the sequence diversity
became the central arena of confrontation in the field of immunology and remained so for fifteen years. Two camps emerged relatively early. In one camp were those that held to the view that

relatively few genes in the germline were acted upon by a series of
somatic processes, each of which contributed to immunoglobulin diversity, while the other camp held that the diversity existed totally
in the germline and had been acted on through evolution; that for
every "new" amino acid sequence that was deducted, a new germline
gene had to be added. The number of genes required from the formal
data was approaching the thousands in the inbred BALB/c mouse and
it was increasingly clear that neither the somatic processes
required to generate the diversity nor the amount of DNA in a germ
cell was sufficient to satisfy the observed phenomena. The ultimate resolution of the dilemma awaited experimental approaches to
the DNA itself.

The immune system can be viewed as a device that plays sentry
over a continuous onslaught of external stimuli and negotiates
with each appropriately for the preservation of the organism. As
a receptor, processor, and interpreter of this information, this
organ system is rivaled by none in sophistication and sensitivity
save possibly the nervous system.

The immune and nervous systems, in fact, share many facets.
Both must respond to a vast array of external stimuli. Both are
comprised of cells which penetrate virtually all other organ systems
and tissues of the body. In consequence, with their responsibility
of preserving the entire organism, both can be conceptualized as
consisting of afferent and efferent modes; i.e., the cells of both
systems not only receive information but transmit it as well.
Critically, both are also modulated by a balance of positive and
negative regulatory mechanisms: excitation and inhibition, depo-
larization and polarization in the case of nerves, help and suppression in the control of lymphocytes. In a sense, the immune
system has an additional level of complexity over that of the
nervous system, though this very complexity also affords immunologists an experimental advantage over neuroscientists. While the
nervous system is well moored in the body in a static web of
axons, dendrites and synapses, the elements of the immune system
are in a continuously mobile phase scouring over and percolating
through the body tissues incessantly, returning through an obscure
system of lymphatic channels and then blending again in the blood.
This dynamism is relieved only by scattered concentrations called
lymphoid organs.

In large part, owing to the accessibility of the immunoglobulin molecule and the ease of manipulation of a cellular system
unencumbered by any rigid organ matrix, the immune system has
served as a forerunner providing a model for differentiation in
the storage and expression of information in a complex biological
system. More than most other systems, it is amenable to detailed
cellular and molecular analysis.

After the discovery of antibodies in serum, Paul Ehrlich was, in 1900, among the first to consider in detail the problem of immunologic specificity. He envisioned immunocytes as pluripotential, each equipped with a diversity of "side chain groups" exposed to the environment of the cell surface. Antigens, on introduction into the organism, were then viewed as interacting in a complementary way with one or more side chain constituents, resulting in the release of that side chain(s) in complex with antigen from the immunocyte cell surface. This release triggered the synthesis of more of that side chain group in the corresponding cell. This construct constituted the first <u>selective</u> <u>theory</u> of antibody diversity, suggesting that the role of antigen was to select a complementary antibody and amplify that antibody with a selective immune response. The ability of an antibody to discriminate between the homologous antigen and other substances results from the complementary nature of the two reacting surfaces. In Ehrlich's words, "the groups . . . must be adapted to one another, e.g., . . . as lock and key," echoing precisely Emil Fischer's earliest sentiment regarding enzymes. Implicit to the concept of preformation of antibodies in Ehrlich's theory is an assumption of the origin of antibody diversity. The problem of antibody diversity reduces to the question of where and how in the antibody lineage antibody diversity is actually deduced. The germline hypothesis supposes that it arises in the germline and that it is propagated or discarded by natural selection to the extent that the germline gene contributes to the overall fitness of the individuals in which it is present. It would follow then that the genetic information for the entire range of antibody specificities which an individual can produce is encoded in the DNA of the zygote, and that diverse structural genes for different antibodies are present in the genome. Thus, since Ehrlich proposed that the antibody repertoire is predetermined, it is to the germline hypothesis of antibody diversity to which Ehrlich's theory implicitly subscribed, spawning the first of the two major camps into which the controversy of the origin of antibody diversity divided.

Elements of Ehrlich's model fell from favor in the 1920s and 1930s as this period brought with it an expansion and characterization of the notion of immune specificity, particularly through the work of Landsteiner which demonstrated that the immune system had an exquisite specificity and a near global range of responses. Haurowitz in 1930 countered the selective theory by applying the following reasoning:

> The unlikelihood of any direct action of the antigen in the organism led Ehrlich to the conclusion that small quantities of antibodies are preformed in the organism and that the injection of antigen merely increases the level of normal antibody synthesis. . . . This hypothesis could be satisfactory as long as only a few natural

antigens were known. Since the work of Landsteiner and
his co-workers, however, an almost unlimited number of
artificial antigens . . . can be produced which have
never appeared in nature. It is unimaginable that an
animal routinely produces predetermined antibodies
against thousands of such synthetic substances.

Rather, it seemed more reasonable to the theorists of the time
that antigen must <u>instruct</u> the immune system in some fashion and
provide its own information for the construction and synthesis of
antibody molecules. These features highlight the embodiment of the
instructionist theories. This must further be regarded as a singular development in theoretical immunology as it ushered in a state
of speculation about alternative theories of diversity predicated
on the apparent limitlessness of the immune repertoire. This construct sought to circumvent the traditional line of natural selection operating on germline genes.

This primordial age of ideas fomented the second major camp in
the antibody diversity issue, collectively constituting the <u>somatic</u>
<u>hypothesis</u> which proposed that antibody diversity arises separately
in each individual. Albeit in continually updated forms, the somatic
and germline hypotheses of the 1930s and continuing through the
present have survived major explosions of information in immunology,
particularly at the molecular level, as the polemical scaffolds
around which the controversy of antibody diversity continued to be
waged.

Linus Pauling in 1940 defined the instructionalists' interpre-
tations in molecular terms by proposing that antigens serve as
templates which permit the antibody polypeptide chain to fold
around it in a complementary fashion and thus mold itself into a
specificity for that antigen. By this interpretation, antibody
diversity would arise by the variety of three-dimensional configu-
rations that these molecules could assume under the influence of
antigens. This theory had immediate difficulty with certain known
properties of the immune response, e.g., immunologic memory. It
required the persistence of antigen in the cell for very long
periods of time. Leaping ahead from this historical junction and
using contemporary hindsight, the template theory can be discounted
as untenable. We know that antibodies, like other proteins, are
synthesized from amino acids utilizing mRNA, not antigen, as its
template. Studies by Anfinsen and others in the early 1960s indicated that the primary amino acid sequence of proteins determines
its three-dimensional configuration, rendering any instructionalist
theory at the level of the polypeptide chain unlikely. Haber and
then Whitney and Tanford in the mid-1960s vitiated the remnants of
the theory by extending the concept directly to antibodies. They
chemically unfolded antibodies of known specificity in strong
denaturing solvents and cleaved the disulfide linkages to remove

all possible structural information except that of the primary amino acid sequence, then renatured the molecules in physiological medium in the absence of antigen to find restoration of the original binding specificity. This result has been reconfirmed several times and the concept has been amply reinforced by amino acid sequence studies of immunoglobulins.

A theory to replace antigen template models was present long before that theory was proved to be inadequate. The first form of a modern selectionist theory of antibody diversity was probably Jerne's who, in 1955, postulated the presence of antibodies of all potential specificities in the serum of each individual. Antigen, when introduced, complexed with its complementary antibody and this complex was subsequently recognized by an immunocyte which, by an unknown mechanism, stimulated the synthesis of that antibody. The seminal elements of this theory were soon expanded into the prototype modern selection theory by Burnet and by Talmadge as the clonal selection hypothesis which has so permeated the field that it is difficult to think of immunology outside of its framework.

The power of the clonal selection theory is concentrated in its clear, economical and uncontrived explanation of the major problems of the immune system: the specificity of antibody production, the secondary response and tolerance to self-antigens. The basic premise of the theory borrows from bacterial genetics. Bacterial variation which results from random mutation and by analogy antibody diversity, are preadaptive; i.e., the capability to respond to an environmental agent is present before that agent is introduced. Burnett proposed that antibody structural genes accumulate random somatic mutations in the course of cell proliferation during the development of the immune apparatus. Each antibody variant (a modern phrasing would be V-region variant) thus arises in a single cell and is carried indefinitely in the progeny of that cell or clone of cells. Although somatic mutation was the assumed origin of antibody diversity, it is not a necessary element of the clonal selection theory. Those clones committed to the production of antibody variants reactive with self-antigens are purged, at or shortly after birth, thus eliminating inappropriate clones and explaining self-tolerance. Exogenous antigen, on the other hand, would influence cells with an antibody reactive to it in a positive fashion resulting in stimulation, proliferation, and production of that antibody. This accounts for the specificity of antibody produced on immunization. The accelerated and enhanced antibody response that is produced in a memory response owes to a prior contact and expansion of these specific cell clones with their antigen in the history of the individual's immune system. These, in turn, should respond in greater force and numbers on secondary challenge. At least three major postulates are implicit in this theory of heritable cellular commitment: (1) the antigen receptor site, presumably on the cell surface, and the antibody

combining site whose synthesis that cell controls are identical and
derive at least partially from the same structural gene, (2) the
condition guaranteeing the correspondence of the immunoglobulin
synthesized with the antigen receptor site is that they are constrained to the same cell; i.e., the cell is specialized for the
synthesis of a single antibody, and (3) the cell specialization
stipulated in number two is inherited and, therefore, clonal.

Sporadic refutations of these postulates have occurred in the
literature but the great bulk of evidence is supportive of the
Clonal Selection Theory. The theory's sheer eloquence, however,
has probably been most responsible for its dominance of thought in
immunology and its acceptance as dogma since the early 1960s.

It has been known for many years that the most striking thing
about an immune serum is the specific nature of its interaction
with antigen. For example, an immune serum can distinguish between
them. It can also discriminate between two chemical compounds
differing in a single functional group, between D and L amino
acids, and between many other closely related compounds. This work
was obviously instigated by Landsteiner and his collaborators and
has been one of the hallmarks of immunologic thinking for fifty
years.

How Extensive is the Repertoire?

The actual extent of diversity has been approached both experimentally and theoretically by a number of investigators. One of
the most widely quoted studies is by Williamson who investigated
the antihapten immune potential of single mice immunized with car-
riers to which the hapten o-nitro-p-iodophenyl (NIP) had been con-
jugated. Williamson developed a clone dilution transfer technique
by which he was able to transfer single clones of cells producing
antibodies to NIP to irradiated mice. The individual NIP-binding
immunoglobulins produced by these clones could then be identified
by the pattern they exhibited after resolution by isoelectric
focusing (IEF). The reasoning was that if one counted the total
number of clones transferred and then identified from the IEF
pattern how many identical or "repeating" clones that were in that
number, it would be possible to infer from statistical considerations how many different anti-NIP clones a single mouse could produce. After considerable analysis, a repeat of only five clones
out of 234 transferred was found. From this, it was calculated
that there was a probability that there existed between 8,000
and 15,000 individual cell clones in a mouse capable of forming
antibodies to the simple hapten, NIP.

The implications of these kinds of data are truly astounding.
If the simple hapten, NIP, can elicit in the neighborhood of
10,000 different antibody molecules, then the repertoire must be

astronomical. Investigators over the years have quoted the often
stated notion that a vertebrate immune response represents well
over 10 million different specificities, i.e., 10 million different
antibody molecules. Early on, it was appreciated that 10 million
genes was likely an impossibility in a strict germline sense.
Since it was known at this time that both chains were needed for
antigen binding, there emerged the so-called p x q hypothesis
which taught that every light chain could combine with every
heavy chain in a relatively random manner. It is important to
appreciate that the hypothesis did not specify that antigen selection would not deviate the immune response such that the majority
of molecules might not show a random distribution of heavy and
light chains, but only that every combination was possible. Thus,
this allowed 1,000 heavy chain genes to combine with 10,000 light
chain genes to provide the appropriate 10,000 x 1,000 = 10,000,000
specificities requiring but 11,000 genes. Obviously, it could
also be done with about 6,000 genes (3,000 heavy and 3,000 light).
It is extremely difficult to pinpoint the place and time at which
this 10 million number of specifies emerged into immunological
thinking. It can be found in the writings of the 1940s and 1950s
and by the 1960s and 1970s it seemed like a "fact."

Thus, the problem in a nutshell was how could 10 million
specificities (or 10,000 different antibody molecules to NIP) arise
in a single mouse.

Somatic or Germline?

The basic theory of evolution, all life from preexisting life,
has as its molecular counterpart all DNA from preexisting DNA. It
is then a direct logical consequence that biological information
has descended from a small amount of ancestral information and is
related to it by a multiplicity of ordered divergences, each of
which in essence represents the replication of a DNA molecule.
The ordered divergencies are what allow the representation of the
progression of alteration in DNA information as a genealogical
tree. The diversity of this information arises as small discrete
steps (mutations), each of which occurs in a single molecule. The
endless proliferation of information is offset by the constant
attrition of information contained in organisms not selected for
reproduction.

In multicellular organisms, the overwhelming preponderance of
the organism is invested in somatic cells, and consequently, most
of the divergences in an organism correspond to cell division
during the life of the individual organism. These divergences or
mutations are called somatic and are only transmitted to some of
the cells of the individual in which they arise. Additionally,
any further manipulation superimposed on the preexisting DNA or
its products whether they occur before, during, or after

transcriptional events, is also designated somatic. A few of the
divergences occur in the germ cells. All of the variations accumulated therein are referred to as germline. Mutations occurring in
the germline can be inherited by the descendants of the individuals
in which they occur. It is between the concepts encompassed by the
terms somatic and germline that immunological thought has traditionally meandered, the central question of antibody diversity has
largely been the relative contributions of each of these concepts
to the antibody problem: How do the specificities arise?

There Are a Modest Number of Germline Genes

There are approximately 200 mouse V-kappa gene segments. This
estimate is based on two kinds of experiments both of which depend
on protein structural information, but which take advantage of DNA
analysis. The first is liquid hybridization analysis. Most of
these studies concluded that there were relatively few genes per
subgroup (between one and four). The second type of study, by
Southern blot analysis, tends to give somewhat higher numbers
when virtually identical types of experiments are done. While
there has been no systematic study of each of the approximately 30
mouse subgroups that have been defined in the V-kappa system by
protein sequence analysis, enough is known so that it is clear
that the number of genes per subgroup can vary considerably. In
some situations (MOPC 167) when a cDNA probe is made, only a
single gene is detected in the BALB/c genome by Southern blot
analysis. In other circumstances, approximately 15-20 bands can
be detected. If we take as an average figure from a number of
different studies that have been done that approximately five
V-kappa gene segments are detected in the genome for each subgroup
and assuming that the number of subgroups in the mouse has not yet
plateaued, but will do so around forty, then we conclude that
there are approximately 5 x 40 or 200 mouse V-kappa gene segments
It is unlikely that this number is off by more than a factor of
two and we anticipate that at most the number will be from one to
four hundred.

There are approximately 50 human V-kappa gene segments. The
most extensive study of the human V-kappa system was done by
Rabbitts who prepared cDNA clones from two different human V-kappa
subgroups, V-kappa-1 and V-kappa 3. Together these subgroups at
the protein level (both serologically and by amino acid sequence
analysis) represent 75 per cent of all the human kappa proteins.
When the two cloned V-kappa genes were studied by Southern filter
hybridization against human genomic DNA, each detected approximately 15-20 genes. Over half of these genes were the same.
These data suggest that there is extensive cross-hybridization
between V-kappa-1 and V-kappa-3 in man (there are whole stretches
of both the protein and the DNA sequences of these two subgroups

that are identical). Since these represent approximately 75 per
cent of human V-kappa sequences, it is anticipated that in man
these represent a very substantial part of the total V-kappa human
gene pool. Since the cloned probes are unable to clearly distinguish the different subgroups, it is unlikely that in man there
are more than fifty germline V-kappa segments which contribute to
antibody diversity.

There are two mouse V-lambda gene segments. Studies by
Tonegawa early demonstrated the presence in the germ line of
V-lambda-1 and V-lambda-2 sequences and related these two sequences
to the known protein sequences of the lambda proteins that had
been sequenced by others. In fact, as mentioned earlier, these
data represented some of the earliest and strongest arguments
against the germline theory. Extensive analysis largely by
Tonegawa's group has shown that there are no additional mouse
V-lambda genes and it is likely that the entire mouse lambda repertoire (approximately 5 per cent of mouse immunoglobulins) derives
from these two V-lambda gene segments.

There are approximately 20 human V-lambda gene segments. The
human V-lambda gene segments have not attracted a great deal of
attention at the present time, although some studies have been
reported. The results are not very different from those reported
above and indicate approximately four genes per subgroup. Since
there are five human V-lambda subgroups defined by protein sequence
sequence analysis, this would suggest that there are approximately
4 x 5 or 20 human V-lambda gene segments.

There are approximately 100 mouse V_H gene segments. These
studies were essentially identical to the studies done in the mouse
V-Kappa system. cDNA copies of the variable regions from myeloma
proteins belonging to particular mouse V_H subgroups were constructed
and studied on mouse embryonic DNA. As few as one and as many as
30 bands have been detected by Southern filter hybridization analysis, although the average seems closer to 10. Extensive amino
acid sequencing of mouse V_H regions would indicate that there are
approximately ten V_H subgroups. Thus, with an average of 10 genes
per subgroup and 10 subgroups, there are approximately 100 mouse
V_H gene segments.

There are approximately 100 human V_H gene segments. Extensive amino acid sequence analysis indicates that human V_H genes
like human V-kappa genes seem to cluster more simply into fewer
subgroups than the mouse. Thus, in man there are approximately
five V_H subgroups rather than the 10 detected in the mouse. Using
a cloned probe for the human V_HIII subgroup, Rabbitts detected
20 hybridizing bands in human placental DNA. If these results are
extrapolated to the other four subgroups, there are approximately
100 human V_H gene segments.

There are three to six J_K gene segments depending on the species. By DNA analysis, the mouse appears to have five J_K gene segments, although one of these is not utilized. In humans, there are five J_K gene segments; in the rat there are six.

There are four to six J-Lambda gene segments depending on the species. In the mouse, there are four J-lambda genes but one is a pseudogene (see below). Humans have six J-lambda gene segments. The arrangement of J and C in both lambda and kappa differs in different species (see below).

The number of J_H gene segments varies from four to six depending on the species. Mouse J_H segments have been detected by DNA analysis to be four. In humans, this number is six.

There are a modest number of D segments, but the exact number is unknown at present. The D segment of both mouse and man represents the least understood region of the human and mouse genomes at the present time. The extent of variation and the number of D segments cannot be precisely quantitated largely because the D segment is so small that accurate Southern blot analysis is virtually impossible. By sequence analysis, approximately 10-20 D segments have been detected by different investigators in the human and mouse. Since the full extent of the D segment remains to be elucidated, the number of D segments is placed at 10 for further analysis but it is appreciated that this is an absolutely lower limit.

The Number of C Region Genes is Limited and Varies Considerably Among the Species

Human and mouse kappa. These represent the simplest systems under study as there is now extensive evidence that there is but a single mouse kappa C region gene and a single human kappa C region gene. Every other system is far more complicated. The known allelic variations in both mouse and human C kappa have been clearly delineated by DNA analysis and confirm protein chemical analysis which suggested that there were relatively few amino acid differences in the allelic forms.

Mouse lambda. The organization of the mouse lambda C region provided the first insights into extensive complexity in C region gene organization. The entire organization of V, J, and C lambda which has been worked out largely by Tonegawa's group is unorthodox; that is, J segments are placed in juxtaposition to C segments. These data confirm at the DNA level the known associations between V segments and C segments which had been studied at the protein level a decade earlier. One of the four mouse C lambda genes has been shown to be a pseudogene (see below). This form of gene organization lends itself to extensive speculation concerning regulation.

Human lambda. Even more complex is the human C lambda locus where a minimum of six C-lambda genes have been detected to date. The organization of the J segments in relationship to these has not been worked out but it is likely to be similar to mouse lambda based on the Fett and Deutsch work of a decade previously.

Pseudogenes

Approximately one-third of the germline genes may be pseudo-genes. So far we have described the presence in the germline of approximately 400 gene segments. When DNA sequence analysis was done on gene segments of both man and mouse, and for both V and J segments, pseudogenes have been detected with remarkable frequency. For example, of three V_K genes sequenced by Rabbitts from a human genomic library, one was a pseudogene and the mouse J-lambda-4, C-lambda-4 structures are also pseudogenes. In mouse, V_H pseudo-genes are also common. From an analysis of the presently available V-kappa, V-lambda, V_H, and J_H gene segments, approximately one-third have been pseudogenes.

There are several ways in which pseudogenes are identified by sequence analysis. They may lack appropriate promoter regions, they may have base substitutions which would introduce amino acids which may never have been seen in that particular position pre-viously, especially in positions that have been known to be highly conserved, for example, tryptophans and cysteines. Finally, and more easily identified, are those genes that contain chain-terminat-ing codons within the reading frame.

Gene Families and Specificity

An important point to stress here is that counting genes by Southern blot hybridization provides an estimate of the number of genes in the genome that are structurally related. When one attempts to relate this to specificity, there are only modest cor-relations that can be made. Thus, for example, most antibody responses to even simple haptens are composed of antibodies that generally fall into families of structures with similar V_H and V_L regions. For example, in the arsonate system, an analysis of a large number of hybridomas indicate that there are 3 gene families that are utilized. cDNA probes from any one of the families detects 20 germline bands, but only some of these overlap. While only one of these bands is used in the idiotype positive immune response to arsonate the other 19 may be either pseudogenes or used in the immune response to other antigens.

Precis

Tables 1 and 2 present the number of germline genes that are likely present in mouse and man. As can be seen from this analysis

Table 1. The number of germline genes in the mouse

| Family | V Gene Segments | | |
	V	D	J
V-kappa	200 (150)	0	5 (4)
V-lambda	2 (2)	0	4 (3)
V_H	100 (75)	10	4

Table 2. The number of germline genes in man

| Family | V Gene Segments | | |
	V	D	J
V-kappa	50 (35)	0	5 (5)
V-lambda	20 (15)	0	6 (?)
V_H	100 (75)	10	6 (6)

Note: Figures in parenthesis reflect approximate (V_K, V_λ, V_H) or exact (J_K, J_H) number of gene segments which are not pseudogenes.

alone, there cannot be enough germline information to encode the required million to billion antibody specificities, idiotypes or hypervariable regions that have been suggested to exist. Clearly additional mechanisms are needed as will be described below.

Combinational Joining of V/J and V, D, and J Gene Segments

Shortly after the discovery of the V and J gene segments in mouse lambda, Weigert's group reanalyzed the protein sequences of the V-kappa-21 subgroup of mouse light chains and proposed that combinatorial joining of V gene segments and J gene segments partially explained antibody variability. Since that time, the principal of combinatorial joining of V/J and V, D, and J gene segments has become more appreciated and is seen to be one of the major mechanisms which amplifies the limited amount of genetic material.

Mouse V-kappa. The mouse V-kappa system perhaps represents the best example of how V and J gene segments can join in a combinatorial fashion to further generate antibody diversity. We have discussed the fact that there are approximately 200 mouse V-kappa gene segments, perhaps 50 of which are pseudogenes, leaving about 150 mouse V-kappa gene segments. Of the 5 mouse J-kappa

gene segments, one is a pseudogene, leaving 4 functional mouse
J-kappa gene segments. Since the evidence suggests that any of the
V-kappa gene segments may combine with any of the J-kappa gene segments, this provides 150 x 4 = 600 totally different complete V_K
genes that can be generated from this limited amount of information.

There is no absolutely firm evidence that every one of these
combinations can occur. Clearly in certain immune responses, particular V_K gene segments are combined with particular J_K gene segments (for example, the arsonate system described earlier). However,
as in the V_K 21 system, different V_K gene segments can combine with
different J_K gene segments greatly amplifying the diversity. Thus,
while there are clear preferences, the evidence seems reasonable
that most combinations appear to occur and occur with a reasonable
randomness suggesting that any mouse V_K gene segment can recombine
with any mouse J_K gene segment.

Mouse V_H. With approximately 100 mouse V_H gene segments,
approximately 25 of which are pseudogenes, we are left with
approximately 75 mouse V_H gene segments (see Table 1). We have
estimated that there are approximately ten mouse D segments, and
there are four functional mouse J_H sequences. The evidence is not
completely in on this particular issue, but it appears that V segments can be found in association with different J_H gene segments,
and certainly the evidence is overwhelming that many different V
segments can combine with the same J segments. Similarly, the evidence is good that there is no general obvious rule to suggest a
preferential association of these except within a particular antibody response. This combination of V, D, and J provides another
measure of variation and provides approximately 3,000 V_H genes from
this limited amount of germline information.

Junctional Diversity Is Created by Alternate Frames of Recombination between Germline V-, and J-region Gene Segments

The extent of junctional diversity has best been described in
the mouse V-kappa system, although studies in mouse V_H are nearly
as complete. The basic mechanism involved in junctional diversity
involves the notion that recombinational flexibility can amplify
diversity by changing the crossover point at which the V and J
sequences recombine. Since this can vary over a range of several
nucleotides, it can give rise to different nucleotide sequences in
the active gene. The result in the light chain, for example, is
that the codon for position 96 can vary dramatically. This high
variability at position 96 was evident a decade earlier in
the first analysis of immunoglobulin light chain protein sequences
by Wu and Kabat who noted that position 96 was the most variable
position in immunoglobulin chains. Since this is a region that is
involved in the antibody combining site and the idiotypic determi-

nants, variability in this region is likely important for the
function of antibody molecules. The extent to which this contributes
to antibody diversity can only be speculated upon. Certainly at the
theoretical level, if one studies a large number of V-kappa gene
segments and J kappa gene segments for every combination, there are
approximately five different amino acids that can be generated by
the flexibility of recombination. (In practice, this may be closer
to four.) In the most thorough analysis of this which was done by
Leder, most of the predicted sequences had been seen.

    In the heavy chain obviously this junctional diversity operates
both in the V_H-D as well as the D-J recombination. And, therefore,
in heavy chains this increases diversity by a factor of 4 x 4 or 16.

    Different systems utilize flexible recombination in different
ways. For example, the arsonate light chain genes of both A/J and
BALB always generate an arginine at position 96 which is present in
neither the germ line V_K nor the germ line J_K genes. Thus, a new
amino acid is generated which is not present in the germ line but
is reproducibly put into the V-kappa gene for function in an anti-
arsonate antibody. Thus, flexible recombination in this system
generates an amino acid and always the same amino acid, but an
amino acid that is not present as a germline encoded
event.

    The arsonate heavy chains display a maximum of variability and
with only a couple of exceptions, every arsonate heavy chain
sequence has a different amino acid at the V-D and D-J junctional
position.

    The theoretical diversity that can be generated by this, then,
represents an additional factor in the light chain, thus 600
(see section 8.2) x 4 = 2,400 V-kappa genes and in the heavy chain,
3,000 x 16 = 48,000 V_H genes. Thus, we see how a limited number of
germline genes can generate almost 50,000 different active genes
by combinatorial and junctional diversity. While this is likely
enough to account for the required diversity seen in the immune
system, it clearly does not satisfy all of the data as there are
three hypervariable regions in each chain not one (note that the
recombination at position 96 and at the V, D, and J junctions only
contribute to the variability one or two positions in at the third
hypervariable region of each chain). In order to account for
variability in the first and second hypervariable region as well
as in the framework, we must turn to even additional mechanisms, as
those are unlikely to be encoded in the germline since there are
enough proteins sequenced through the first hypervariable region
of both the heavy and light chain to indicate that there is not
enough germline information to account for this variability.

Somatic Mutation in All Gene Segments Leads
to Further Diversity

The earliest positive direct evidence for the somatic theory
of antibody diversity that has been adequately described up to this
point comes from the mouse lambda system where Tonegawa early demonstrated the presence of a single lambda variable region gene for
each of the two lambda V region subgroups. Since there were nine
known lambda-1 and one lambda-2 proteins that had been sequenced,
a direct comparison of the two indicated that somatic mutation had
occurred in the eight variant V-lambda-1 proteins. This would
indicate that each germline gene could undergo a minimum of eight
somatic variants which could derive from a single germline gene.
Statistically, however, considering the number of lambda proteins
that had been sequenced as we had described before, there were,
with better than 90 per cent confidence limits, over 16 different
lambda variants expected. This has suggested to many investigators
that in the lambda system at least 20 variants can be derived from
a single germline gene.

In the heavy chain, Hood's group has analyzed in extreme detail
the immune response to phosphorylcholine at both the protein and
nucleic acid level. Amino acid sequence analysis of the heavy chain
variable regions of 19 hybridoma and myeloma immunoglobulins were
available. Ten of these were identical to the prototype T15 and
nine were distinct variants differing by one to eight residues. A
cDNA probe was made to the T15 protein. Using the T15 V_H DNA probe,
Hood and his group were able to isolate four closely related members
of the T15 V_H gene family including one that encoded the T15 V_H
sequence itself. A comparison of the germline sequence with the
protein sequences indicated that the entire immune response to
phosphorylcholine derived from a single T15 single germline gene.
The somatic variation that they detected in the expressed genes
was extensive in and around the coding regions of the two variant
V_H genes and that it was found in flanking as well as coding
sequences. Many of the substitutions were silent. These data provided firm evidence that somatic mutation occurred in the V gene
segment.

Similar studies have been done in the arsonate system where it
has been established that a single germline V_H gene gives rise to
the entire repertoire of arsonate antibody that bear the cross-
reacting idiotype. Since over 35 of these molecules have been
sequenced and essentially all are different, since they derive
from a single germline gene, upwards of 35 variants are possible.

It is likely that the extent of the amino acid sequence diversity that result from somatic mutation is even higher than the
three figures that have been presented here: 8 for V-lambda, 9 for
T15, 35 for arsonate. Since none of the systems have been exploited

to a plateau, conservatively, however, it seems clear that each germline V gene segment can give rise to a minimum of 10 variants. This allows a further permutation of the numbers developed previously to suggest that in the mouse V-kappa system, 2,400 x 10 = 24,000 or 2.4 x 10^4 complete V_K genes, and in the heavy chain, 48,000 x 10 = 480,000, or 4.8 x 10^5, complete V_H genes.

In addition to somatic variation in the V gene segments, somatic mutation has been documented in the J segments as well. In the arsonate system in both the heavy and light chain, there are J segment variants.

In the arsonate heavy chain J segments sequenced, variations are relatively common. These data indicate that J segment somatic mutations are relatively common. It is fair to say that other systems do not seem quite this variable and we will not further increase diversity based on this mechanism until more information is available. It is likely, however, that this leads to further diversity.

The importance of the J segment in antibody diversity was shown by Scharff and co-workers who isolated variants of the S107 myeloma protein and showed that variations in the J gene segment could lead to an alteration of binding activity for phosphorycholine

Combinatorial Pairing of Heavy and Light Chains Further Amplifies Diversity

There is no formal evidence that every light chain can pair with every heavy chain. However, the large bulk of experimental data suggests that some form of random combinatorial pairing of heavy and light chains can be utilized to amplify antibody diversity. There is abundant evidence that different light chains, for example, can compare with the same heavy chain. This is particularly evident in the antiphosphorylcholine antibodies of the BALB/c

Table 3. Contributions of germline, somatic and interactional processes to murine antibody diversity

	Germline			Combinatorial	Junction (X4)/Junction	Somatic Mutation (X10)
V_K	$150V_K$	$4J_K$		600	2,400	2.4 x 10^4
V_H	$75V_H$	$10D_H$	$4J_H$	3,000	48,000	4.8 x 10^5

antiphosphorylcholine antibodies of the BALB/c mouse where the same
T15 gene is used to generate the heavy chains, but three distinct
light chain genes are used to generate three different families of
anti-PC antibodies. Thus, in this example, three light chains can
recombine with the same heavy chain. In other systems, multiple
heavy chains have been found in association with the same light
chain. Using the extreme of the argument has been pointed out
earlier. The random association of a 1,000 light chains and 1,000
heavy chains would produce 10^6 different antibodies. Referring to
Table 3, the random association of 2.4×10^4 light chains and
1.2×10^5 heavy chains (leaving aside lambda chain contributions)
could generate over 1×10^9 antibodies.

SUMMARY

Thus, we see a limited number of germline genes through the
processes of combinatorial joining of various gene segments of
junctional diversity at the points that gene segments recombine,
that somatic mutation and potentially gene conversion, and finally
the combinatorial pairing of heavy and light chains can generate
an enormous amount of diversity in antibodies. The evidence presented argue strongly that all of these processes contribute
significantly to antibody diversity and that antibody specificities, hypervariable regions and idiotypes can be generated from
a limited amount of genetic material by a whole assortment of
mechanisms that the immune system utilizes to store, manipulate,
and retrieve genetic information.

DISCUSSION

FOX: Are you saying that the huge number of antibodies can simply be explained essentially by assembly processes from a much smaller number of synthetic products?

CAPRA: Right. But on top of that, there's a fairly good load of somatic mutation.

TOBACH: Have you anything to say about the work at the Salk Institute with regards to this system within a neuron? Have you followed that?

CAPRA: This system within the neuron?

TOBACH: Yes. Apparently this concept is a very powerful one in terms of looking at the molecular characteristics of what's going on in the neuron, making new assemblies from new materials for something that was never in existence before and it's a very powerful idea and I'd just like to see it.

MONROY: It's in a book by Edelman and Mountcastle, The Mindful Brain.

KOSHLAND: It's really based on work from Colorado originally and I know where they found a lot of extra mRNA in nerve tissue and so a postulate is that the antibody process is being used to generate diversity in neural tissue.

CAPRA: The immune system and the nervous system have often been compared because they have so many similarities in that they both require an afferent and an efferent limb. They both excite and repress; they both have a stimulus and a response, and they both have to process a lot of information.

TWINS REARED TOGETHER AND APART: WHAT THEY

TELL US ABOUT HUMAN DIVERSITY

Thomas J. Bouchard, Jr.

Psychology Department
University of Minnesota
Minneapolis, Minnesota

The exceedingly close resemblance attributed to
twins has been the subject of many novels and plays,
and most persons have felt a desire to know upon what
basis of truth those works of fiction may rest. But
twins have many other claims to attention, one of which
will be discussed in the present memoir. It is, that
their history affords means of distinguishing between
the effects of tendencies received at birth, and of
these that were imposed by the circumstances of their
after lives; in other words, between the effects of
nature and of nurture. (Galton, 1875, p. 4661)

With these words, Galton began a tradition of behavioral
research with twins that has continued uninterrupted to this day.
In spite of a small sample size, Galton drew some strong conclusions from his work with twins.

There is no escape from the conclusion that nature pre-
vails enormously over nurture when the differences of
nurture do not exceed what is commonly found among
persons of the same rank of society and in the same
country. My only fear is that my evidence seems to
prove too much and may be discredited on that account,
as it seems contrary to all experience that nurture
should go for so little. (p. 576)

Galton's studies were based on a faulty conceptualization of
the phenomenon of twinning (Price, 1950), but both the broad general conclusions and the recognition of probable resistance to
them were reasonably accurate.

147

In spite of their early origins, twin research designs are an extremely underutilized resource in the behavioral and social sciences. The reasons for the neglect of this methodology are complex and beyond the scope of this paper. Suffice it to say that they reflect, in part, the political and methodological biases of the American social and behavioral sciences, namely a preference for environmental explanations of variation and the dominance of the experimental method (cf. Cronbach, 1957, 1975).

In this paper I will describe the principal twin designs and discuss some of their advantages and disadvantages. I will also describe in some detail the Minnesota Study of Twins Reared Apart. Finally, I will present a sample of findings that the designs have yielded in the domains of physical measures, intelligence, and personality.

THE STUDY OF TWINS

The Classic Twin Design

The classic twin design consists of the comparison of monozygotic and dizygotic twins reared together. They may be compared on dichotomous traits by using a variety of measures of concordance or they may be compared on continuous traits using variances or intraclass correlation coefficients. Discussion will be restricted to the continuous case.

The measured value of a continuously distributed character (e.g., height, or IQ) in a given individual is called a phenotypic value (P). This value is determined by both genetic (G) and environmental (E) factors. The three components can be expressed as follows:

$$P = G + E$$

The phenotypic value is composed of the genotypic value and the environmental value. Across a population of individuals these components have means and variances. Under appropriate assumptions the variances, but not the means, are additive. Thus:

$$V_p = V_g + V_e$$

This formulation requires the assumption of independence between the genotypic and environmental values. If the genotype and a feature of the trait-relevant environment are correlated, we must include a covariance component.

$$V_p = V_g + V_e + 2COV_{ge}$$

The genetic variance (V_g) can be divided into additional components. The principal ones are additive variance (V_a) and dominance variance (V_d). If different loci interact, we have epistasis and must add the appropriate term (V_{ep}). If there is assortative mating for the trait, we must add assortative mating variance (V_{am}). If different environments have different effects on different genotypes, an additional term characterizing the interaction (V_i) must also be introduced. The equation may now be written:

$$V_p = V_a + V_{am} + V_d + V_{ep} + V_e + V_i + 2COV_{ge}$$

To keep the treatment simple we will ignore measurement error.

The narrow heritability is defined as the ratio of additive variance to phenotypic variance:

$$h^2 (narrow) = (V_a + V_{am})/V_p$$

The broad heritability is defined as the ratio of total genetic variance to phenotypic variance:

$$h^2 (broad) = (V_a + V_{am} + V_d + V_{ep})/V_p$$

Notice that we can define a population statistic called environmentality:

$$e^2 = V_e / V_p$$

In terms of Mendelian components, the variances for twins reared together can be expressed as follows:

$$V_{mzt} = V_a + V_{am} + V_d + V_{epmzt} + V_{imzt} + 2COV_{gemzt} + V_{ec}$$

$$V_{dzt} = 1/2V_a + V_{am} + 1/4V_d + V_{epdzt} + V_{idzt} + 2COV_{gedzt} + V_{ec}$$

The term (V_{ec}) refers to the common environment component. Notice that the epistatic, interaction, and covariance terms differ for the two types of twins.

These variances become intraclass correlations when divided by the phenotypic variance.

The most widely used formula for estimating heritability using twin intraclass correlations was introduced by Falconer (1960):

$$h^2 = 2(r_{MZT} - r_{DZT})$$

This formula assumes random mating, no dominance, no epistasis, no interaction, no COVge, and equal common environmental variances for the two types of twins. Depending on the circumstances, then, this formula may give biased results. It is possible to make corrections for assortative mating, but the correction itself depends on additional assumptions. The classical twin method simply does not allow us to test hypotheses about specific genetic mechanisms. From the point of view of a geneticist, this is a disaster. From the point of view of a psychologist, however, things do not look as bad. Virtually no psychological theory specifies environmental effects in any but the crudest fashion (McAskie and Clarke, 1976; Willerman, 1979). The theories that do specify effects focus almost exclusively on the family (Rowe and Plomin, 1981; Scarr, 1982). Consequently, the twin method supplies a quick and easy method for screening traits in terms of their susceptibility to the major classes of environmental effect, namely, within and between family influences.

Over the last few years, Lindon Eaves, David Fulker, and their colleagues have developed a model-fitting approach to family data that allows us to engage in this kind of screening in a fashion that is far superior to the simple calculation of heritabilities. Specifically, it allows us to make precise tests of a variety of competing hypotheses (cf. Eaves and Young, 1981). Using ordinary twin data, the procedure simply consists of fitting observed mean squares to expected mean squares specified by the various theories. We will illustrate the procedure with an example below. One of the nice features of this approach is that it can be systematically extended to include independent data sets gathered on singletons. This enables us to systematically test the generality of inferences based on twins using other data sets.

Twins Reared Apart

The twins-reared-apart design is in principle the simplest of all the twin designs. If there is no placement bias (twins are assigned to homes randomly with respect to trait-relevant environmental factors) then the intraclass correlation between MZ twins is a direct estimate of heritability. More importantly, it has a much smaller sampling error than the heritability estimate derived from ordinary twins. My colleagues David Lykken, Seymour Geisser, and Auke Tellegen have recently shown that an intraclass correlation of .40 based on 50 monozygotic twins reared apart (MZA) has a 98 percent confidence interval of from .13 to .65. To achieve the same confidence interval with ordinary twins, about 800 pairs are required. The correlation between dizygotic twins reared apart (DZA) estimates one-half the heritability.

In practice, studies of twins reared apart suffer from a number of difficulties. Separation of twins is a rare event. Also a group of such twins may not constitute a representative sample of genotypes. The twins who are willing to participate in such a study are to a large extent, a self-selected population and this further exacerbates the sampling problem. Placement of these children in adoptive homes is also unlikely to be random.

The impact of these various biases can to some extent be assessed in order to determine whether they significantly affect conclusions that might be drawn from the results. It is possible, for example, to examine the correlation between trait-relevant features of the environments of the twins and the correlations between these features and the relevant traits, in order to determine to what extent placement has influenced any measure of similarity. The distribution of the twins across such variables as social class and educational level of rearing parents will reveal whether the sample deviates in a significant way from a random sample. The distribution of the scores on the trait of interest is also informative. If the distribution is different in important ways from the distributions of comparable singleton populations, then there are grounds for limiting generalizations based on the twin data alone.

The twins-reared-apart design is a variant of the adoption design. The purpose of most adoption designs (Rosenthal, 1970) is to test environmental effects on traits of interest (IQ, schizophrenia, etc.). We can ask, for example, how similar in IQ are adopted children to their adopting parents. If the necessary assumptions are met then the correlation is a direct estimate of the effect of common family environment on the trait of interest. Clearly it would be desirable to study the adoptive parents of twins reared apart and obtain such information. Unfortunately, this has never been done. Since the differences between monozygotic twins reared apart are entirely environmental in origin, it is possible to correlate these differences with environmental differences in order to assess the influence of specific environmental factors. This can, however, be a tricky matter. In their sample of nineteen pairs of monozygotic twins reared apart, Newman et al. (1937) correlated differences in Stanford-Binet IQ with rated differences in amount of education. They found a correlation of .791. The correlation between educational difference ratings and Stanford Educational Age differences was even higher, .908. This evidence strongly suggests that amount of schooling has a direct influence on IQ. It is possible that amount of schooling is in part a reflection of IQ. The twins could have differed in IQ due to prenatal influences and difficulties in school may have led to earlier termination for the less capable twin. While the effect is still environmental, its specific source is misidentified.

The phenotypic variance of twins reared apart in uncorrelated environments may or may not contain genotype X environment covariances, depending on the trait and one's theory of environmental action (cf. Eaves et al., 1977; Plomin et al., 1977). It does not contain a common environmental component. The interaction terms are included.

$$Vmza = V_a + V_{am} + V_d + V_{epmz} + V_{imza}$$

$$Fdza = 1/2V_a + V_{am} + 1/4V_d + V_{epdz} + V_{idza}$$

The problems of estimating the effects of genotype X environment covariance and interaction have yet to be completely resolved. Some investigators believe they are insoluble (Feldman and Lewontin, 1975).

Plomin et al. (1977) have suggested it might be useful to categorize genotype X environment interactions into three types: active, passive, and reactive. An active genotype X environment correlation occurs when an organism seeks out features of the environment related to his genetic propensities. "For example, bright children may seek peers, adults, or inanimate aspects of their environments that foster their cognitive growth" (p. 310). Passive genotype X environment correlations occur when the environment, for whatever reason, contains features that are related to a developing organism's genetic potential. The most commonly cited human example is that of intelligent parents providing an environment conducive to the development of intelligence in their children, who carry alleles favorable to the development of intelligence. Parents at the low end of the distribution of intelligence would, under this model, provide an environment unfavorable to the development of the intelligence of their children, who carry fewer alleles for the trait. Eaves et al. (1977) consider this covariance due to cultural transmission. Passive covariance is essentially zero in an adoption design. If passive covariance is important in the within-family setting, data from adoption designs should yield reduced variances in comparison to ordinary families. Reactive genotype X environment correlations occur whenever people "respond to genotypic differences among individuals in such a way that they provide an environment that reflects and correlates with those genotypic differences" (Plomin et al., 1977, p. 310).

It has been argued that reactive and active genotype-environment correlation may increase the covariance between identical twins reared apart and thus lead to an overestimate of the genetic variance in such studies. This is true or false depending on how one conceives of the problem. Roberts (1967) has argued:

> The genotype may influence the phenotype either by
> means of biochemical or other processes, labeled for
> convenience as "development," or by means of influencing
> the animal's choice of environment. But this second

pathway, just as much as the first, is a genetic one; formally it matters not one whit whether the effects of the genes are mediated through the external environment or directly through, say, the ribosomes. (p. 218)

As Plomin et al. (1977) point out, from the perspective of behavioral manipulation, it matters a great deal. While the range of environments from which an organism can choose is manipulable, tinkering with ribosomes is not likely to become common practice in the near future. From the point of view of evolutionary theory, however, the distinction is irrelevant. An organism that is more capable of acquiring food is more likely to survive than its competitors, regardless of whether this capacity is mediated through a slightly different choice of ecological niches or a change in muscular structure. Either change results in a correlation between a feature of the genotype and a feature of the environment (amount of food consumed).

The study of MZA twins is an important means of validating the classic twin method. Shields (1978) has pointed out that such twins:

. . . can be used to test the hypothesis that MZTs are more alike than DZTs on account of the more similar within-family environment of the MZTs. . . . For this purpose it is no disadvantage that the environments of the two twins may not differ very much. Indeed, it would be best if the MZA twins were exposed to environments as similar in as many respects as possible, except for having different rearing parents and no close mutual contact. (pp. 91-92)

SOME PROBLEMS AND COMPLEXITIES IN TWIN RESEARCH

The Problem of Scales of Measurement

A number of critics of behavioral genetic research argue implicitly, if they do not state explicitly, that the IQ scale is arbitrary and that it is unlikely that IQ is a trait with any biological significance. Others argue that the IQ scale does not have measurement properties which are adequate to sustain a quantitative genetic analysis. These arguments can be and have been applied to psychological scales in general.

None of these arguments are very convincing nor are they sufficient to disqualify psychological scales from being used for genetic analysis. First of all, the choice of trait and measurement technique for any kind of research should be made on a theoretical and practical basis. An investigator's theory determines

whether or not a particular scale has meaning. Whether a trait has
biological significance is a matter that must be determined empiri-
cally, not a priori.

One of the reasons that this problem arises is that some
investigators believe that there is a "real" scale for each psychological attribute and it is imperative to discover the true scale.
Under this view, once the true scale is known, further work can
proceed. I prefer the view which asserts that scale development
is a fundamental part of theory development and that both go on in
the context of empirical investigation. Lord and Novick (1968),
for example, assert that:

> If a particular interval scale is shown empirically
> to provide the basis of an accurately predictive and use-
> fully descriptive model, then it is a good scale and
> further theoretical developments might profitably be
> based on it. Thus measurement (or scaling) is a funda-
> mental part of the process of theory construction. (p. 22)

Ghiselli et al. (1982) have argued:

> The process of defining our variables and devising
> operations that we can use in the description of individ-
> ual differences is a never-ending one. Not only is there
> an interaction between the definition and the devising
> of operations, but also the results obtained from our
> operations give us new insights into the nature of our
> variable so that we redefine it and modify our opera-
> tions. The development of ways for describing individ-
> uals, then, is a dynamic process. (p. 15)

A behavior genetic analysis can be an important component in the
construction of a theory about a trait even though the trait has
no genetic basis whatsoever.

Finally, it is a widely-recognized but seldom stated principle
that if a trait is poorly measured it will generally give quite
unsatisfactory results in an empirical investigation. Psychological scales are poorly characterized from the point of view of measurement models, but as Thomas (1982) put it,

> This situation does not mean . . . that such scales
> . . . are useless. Obviously the long history of use
> of the Stanford Binet in educational settings simply
> illustrates that tests may be quite useful even when
> their scale properties are poorly understood. (p. 202)

Primary Biases

Primary biases have been defined by Price (1950) "as prenatal and natal environmental factors peculiar in kind or degree to twins" (p. 295). The principal factors are natal factors, lateral inversions, and the effects of mutual circulation. On average, all of these factors occur more frequently among MZ than DZ twins. To the extent that they have an influence on the trait under consideration they tend to make MZ twins more different from each other than they would be otherwise.

Natal factors include type of placentation (the primary distinction being between dichorial and monochorial cases), crowding in the womb, poor intrauterine position (probably only important during the later stages of pregnancy), and conditions of delivery (primarily short gestation).

Lateral inversions refer to the varying degrees of "asymmetrical reversals" in physical structures observed in twins. The most extreme example is that of 'situs inversus viscerum' where the twins show complete transposition of the viscera. The more widely-known phenomenon of mirror imaging involves skin and hair patterns.

Mutual circulation refers to the fact that in monochorional fetuses there is shared blood circulation. Circulation is seldom perfectly balanced and in some cases there is an extreme imbalance. Since a condition of imbalance will be maintained over time the consequences for development can be considerable. A more serious condition called an arteriovenous anastomosis or "transfusion syndrome" causes chronic malnutrition and reduced hemoglobin and serum protein in the donor twin. Twins suffering from this syndrome, have been known to differ in birth weight by over 1,000 grams (typical twin birth weight is about 2,600 grams).

Price (1950) began his classic review of this literature with the following statement: "In all probability the net effect of most twin studies has been underestimation of the significance of heredity in the medical and behavioral sciences" (p. 293). We still have very little knowledge of how seriously these biases may affect psychological research. A recent report by Rose et al. (1981) suggests they may be sizeable for some traits. These authors compared Vocabulary and Block Design scores of adult monochorionic-MZ, dichorionic-MZ, and DZ twins. For Vocabulary the intraclass correlations were .949 (n=17 pairs), .948 (n=15 pairs), and .554 (n=28 pairs). The two chorion types do not differ although the correlations are a bit high. For Block Design the intraclass correlations were .919, .475, and .444. The dichorionic-MZ twins did not differ significantly from the DZ twins. These results "comprise the first evidence of a lasting influence of placental variation on adult intellectual performance" (p. 39).

Given the small sample, these interesting results must be replicated before they can be taken very seriously.

Social Biases

Perhaps the most common criticism of the ordinary twin method is that MZ twins are treated much more alike than DZ twins. It is a short step from this facile argument to the conclusion that any excess similarity among the MZs in contrast to the DZs is thus explained. What is required to vitiate the ordinary twin method is evidence that the environments of the two types of twins differ on trait-relevant dimensions. More specifically, it must be shown that the environmental or treatment variable on which the twins differ is in fact a variable that influences the trait under consideration. Furthermore, the direction of effect must be in the predicted direction. To my knowledge this has never been shown in the domain of mental ability, personality, or psychological interests. Simply pointing out idiosyncratic correlations within various studies (cf. Kamin, 1974; Kamin in Eysenk and Kamin, 1981) proves very little if anything.

On the other hand, numerous predictions based on the unequal environment hypothesis have not been confirmed when subjected to empirical tests. Loehlin and Nichols (1976), for example, attempted to relate differences within pairs of twins to differences in environmental treatments. Consider the oft-cited argument that because many more identical than fraternal twins are dressed alike, the identical twins are more alike in personality. Parents of twins were asked "Were the twins dressed alike?" They responded on a three-point scale; Almost always - 1; Part of the time - 2; Rarely or never - 3. The average correlation between parental response to the item and absolute differences on the eighteen California Psychological Inventory Scales for 451 MZ twins was .004. The composite sum of six items dealing with differential experience has an average correlation of .056 with the absolute differences on the same eighteen scales. Having been dressed alike certainly did not make these twins much more similar.

The so-called "effects of twinness" are also often cited as a serious flaw in the twin design.

> When partners are raised together they begin an exceedingly complex and idiosyncratic interaction known as twinning. Partners seesaw, both by observation and self-report, between close identification with and exaggerated independence of each other. (Farber, 1981, pp. 5-6)

As Rose (1982) has pointed out, this is an old issue settled long ago. There is no evidence that an ubiquitous twinning phenomenon jeopardizes the generalizability of inferences that might be drawn from twin studies.

Another mechanism sometimes invoked to discredit the twin method is that of assimilation and contrast. By this hypothesis, initially similar twins (mostly MZs) are expected to identify with each other more frequently and, over time, come to resemble each other more strongly. Initially dissimilar twins on the other hand would tend to contrast themselves and over time come to be more dissimilar. This latter hypothesis has an empirical consequence that makes it subject to test. If there is a powerful assimilation-contrast effect it should affect the distribution of twin differences and drive it toward bimodality with an excess of small and large pair differences. Loehlin and Nichols (1976) were unable to find any evidence for such an effect in the large sample of twin data that they examined.

Numerous other biases of this type have been postulated and found wanting (DeFries and Plomin, 1978; Plomin et al., 1976; Scarr, 1968; Scarr and Carter-Saltzman, 1979; Scarr et al., 1980). The point to be made here is not that there are no biases, but that postulated biases must be subjected to empirical test. Hypothesized biases are more often than not simply competing explanations of a phenomenon proposed as an alternative to unpalatable findings. Rather than putting them forward as biases, their adherents should put them to an empirical test and either confirm or refute them.

THE MINNESOTA STUDY OF TWINS REARED APART

The Minnesota Study of Twins Reared Apart involves the psychological and medical assessment of identical and fraternal twins reared apart from early in life. Since March of 1979 we have assessed 27 sets of MZA twins, one set of MZA triplets, and 12 sets of same-sex DZA twins. The assessment is carried out by an interdisciplinary team of psychologists, psychiatrists, allergists, cardiologists, and a chronobiologist. The components of the medical assessment are listed in Table 1. Not all of these tests have been carried out on all twins. Items recently added to the assessment are indicated.

Whenever possible, spouses are invited to participate in the psychological portion of the study. Twenty-one sets of spouses have participated to date. The psychological inventories, mental ability tests, and interviews administered to the twins and spouses are listed in Table 2.

Twins are accompanied by a research assistant at all times during the assessment. Many of the inventories and questionnaires

Table 1.   Components of medical assessment used in the Minnesota
Study of Twins Reared Apart

1.   Medical Life History: This includes a review of the complete
history including childhood illnesses, surgery, psychiatric
problems, and a review of systems.

2.   General Physical Examination: A brief general physical exam
is carried out.

3.   The Diagnostic Interview Schedule is administered to each twin
by a separate examiner. (Recent addition)

4.   Sexual Life History Interview and Questionnaires.

5.   Cardiovascular Examination:
a)   Physical Exam by a cardiovascular specialist.
b)   Chest X-ray.
c)   Vector Cardiogram.
d)   EKG.
e)   24-hr. EKG using a holter monitor.*
f)   Overnight blood pressure monitoring. (Recent addition)*
g)   48-hr. body temperature and activity monitoring with a
holter monitor. (Recent addition)*
h)   Stress EKG.

6.   Allergy Testing:
a)   Skin testing.
b)   History of exposure to allergens.
c)   IgE.
d)   Renal chemistry battery.
e)   Serological test for allergens (RAST).

7.   Basic Laboratory Tests:
a)   Complete hematological blood count.
b)   Renal chemistry battery.
c)   Urinalysis.

8.   Zygosity Determination based on 20+ markers and anthro-
pometrics.

9.   Anthropometric Measures:
a)   Height.*
b)   Weight.*
c)   Eye color.
d)   Ear shape.
e)   Finger prints.*
f)   Palm prints.*
g)   Head width.
h)   Head length.
i)   Photographs.*

Table 1 -- Continued

10. Eye Examination:
     a) Brief background medical history.
     b) Visual acuity.
     c) Refraction.
     d) Ocular pressure.
     e) Eye dilation and retinal photographs.
     f) Fundus eye examination by physician.

11. Dental Examination:
     a) Complete intra- and extra-oral-facial exam.
     b) Evaluation of teeth present and missing.
     c) Status of TP (restored, decay, morphology).
     d) Complete radiographic survey.
     e) Study models made.

12. Collection of all previous available medical records when
     possible.

_____

Note: * indicates measures taken on spouses also.

are completed while the twins wait for medical tests and examina-
tions. This is the most comprehensive assessment of reared-apart
twins ever carried out.

     The quality of our sample can be judged from the data in
Table 3 where the attributes of our participants are compared with
the three previous studies.

     As Table 3 shows, the number of twins reared apart and reported
on in the scientific literature is very small. The last cases
reported in the United States were four cases studied by Burks and
Roe (1949). Prior to that, nineteen cases were reported by Newman
et al. (1937). Juel-Nielsen (1965) reported twelve Danish cases
and Shields (1962) reported forty-four British cases. Farber
(1981) has summarized the literature on all MZA twins including a
number of miscellaneous cases. Relative to the three major studies,
our cases were separated earlier and spent more time apart. Our
twins were reunited at twice the age of the Newman et al. and
Shields studies and almost eight years later than the Juel-Nielsen
study.

     It is worth mentioning at this point that the largest study
of MZA twins (53 cases) purportedly conducted by Sir Cyril Burt
has been convincingly shown to be fraudulent (Hernshaw, 1979).
This fraud was uncovered in large part because of the detective
work of Leon Kamin, an ardent antihereditarian (Kamin, 1974).

Table 2.   Inventories, mental ability tests and interviews administered during the assessment

Personality Inventories

Adjective Check List; Activity Preference Questionnaire; California
Psychological Inventory; Differential Personality Questionnaire;
Minnesota Multiphasic Personality Inventory; Myers-Briggs Type Indi-
cator; Fear Survey; Sixteen Personality Factor Questionnaire.

Interest and Value Inventories

Strong Campbell Interest Inventory; Jackson Vocational Interest
Survey; Rokeach Value Survey; Musical Interests; Alport-Vernon-
Lindzey Study of Values; Recreational Interest Inventory; Minnesota
Importance Questionnaire.

Psychomotor Tests

Hole Steadiness Test (involuntary hand movement); Purdue Pegboard
Test; Rotary Pursuit Test (2 days, 25 Tapping Test trials/day).
(The psychomotor tests are not completed by spouses.)

Mental Abilities

Wechsler Adult Intelligence Scale; Raven Progressive Matrices;
Mill-Hill Vocabulary (modified); Verbal Comprehension; Number
Facility; Spatial Orientation - 2 dimensions; Spatial Orientation -
3 dimensions; Spatial Visualization; Speed of Closure; Flexibility
of Closure; Perceptual Speed and Accuracy; Induction (Reasoning);
Word Fluency; Ideational Fluency; Flexibility of Use; Meaningful
Memory; Memory Span; Immediate Visual Memory; Delayed Visual
Memory; Associative Memory; Spatial Scanning; Mechanical Ability;
Spelling.   (Raven and Mill-Hill not completed by spouses.)

Information Processing Measures

Shepard Metzler Reaction Time Paradigm; Posner Reaction Time Task;
Sternberg Task. (Information processing tests are not completed
by spouses.)

Life History Interviews and Questionnaires

Life History Interview; Clinical Interview; Sexual Life History
Interview; Life Stress Interview; Child Rearing/Schooling interview; Briggs Life History Questionnaire; Family Environment Scale.
(Clinical interview, Sexual Life History, and Life Stress Interview not completed by spouses.)

Table 2 -- Continued

Miscellaneous Tests

Diet Questionnaire; Handedness Questionnaire; Smoking Questionnaire;
Television and Reading Questionaire; Requested Facial Action Test;
Barron-Welsh Art Scale.

Table 3. Descriptive characteristics of the participants in the
four major studies of Monozygotic Twins Reared Apart

Studies	Age When Studied (years)	Age at Separation (months)	Age at Reunion (years)	Number of MZ Pairs
Newman et al. (1937)	26.1	1.6	12.5	19
Shields (1962)	28.8	1.4	11.0	44
Juel-Nielsen (1965)	51.4	1.5	15.9	12
Minnesota Study (1982)	36.2	.3	23.9	30

Note. Triplets entered into all calculations three times.

A SAMPLE OF FINDINGS

Physical Similarities

Most people are surprised at the tremendous morphological similarity of identical twins both at a single point in time and across
the entire age spectrum, but they are not astonished. Physical
features are expected to show the effects of heredity.

Fig. 1 shows a set of MZA twins and a set of DZA twins. As
with ordinary twins, the MZs look very much alike and the DZs,
while showing family resemblance, do not look alike. The identical
twins in the photograph turned up at their first reunion wearing
very similar eyeglasses, beards, trousers, and jackets. We were
aware of their impending meeting and were able to bring them and
their wives to Minneapolis within days of their reunion.

Students of twins have always been struck by the tremendous
similarity in the posture and expressive movements of identical
twins. Fig. 2 shows a large group of twins posing for a photograph.
Without being told to do so, every pair of twins has arranged their
hands in the same manner, but in a manner different from every other

Fig. 1. A pair of dizygotic twins reared apart (left) and a pair of monozygotic twins reared apart (right).

Fig. 2. Concordance in identical twins. The physical similarity of the twins, being genetic, is expected, and so is the similarity of dress, which is environmental. The striking feature to this photograph, however, is something else--without having been given any instruction about how to hold their hands, each twin pair has unconsciously put them in a characteristic position. (Courtesy of K. Fredga, University of Lund). From E. Novitski, Human Genetics. New York: Macmillan, 1977, p. 293.

pair. Behavioral traits of this sort can easily be learned by imi-
tation; consequently, such displays do no more than suggest a genetic
basis for gestures. The case reports of MZA twins, however, contain
repeated references to similarities in posture and gestures (Farber,
1981). In our own work, we have repeatedly observed striking similarities in such behaviors as gait, posture, gesticulation, and
various nervous habits such as straightening eyeglasses. Fig. 3
shows a series of photographs taken of MZA and DZA twins at the
beginning of their week at Minnesota. The twins are simply asked
to stand with their backs to the wall. The MZA twins seldom notice
that they tend to assume the same posture and hold their hands in
the same way. Of course not all MZA twins hold their hands in
exactly the same way and some DZA twins do stand and hold their
hands in the same way. We also find striking similarity in the
mannerisms of MZA twins during the videotaped interviews. We will
subject that data to quantitative analysis in the future.

Table 4 reports the correlations between various types of
twins for a variety of physical parameters.

The MZT and DZT data has been collected by my colleagues,
David Lykken and Auke Tellegen, over the course of the last ten
years (cf. Lykken et al., 1974). The ridgecount and height data
fit an additive polygenic model almost perfectly. The DZ twins are
just about half as similar as the MZ twins. The Falconer heritability for fingerprint ridgecount would be 1.00 and for height it
would be .86. These figures are quite close to those typically
reported in the twin literature. They are also very close to the
MZA correlation which is itself an estimator of heritability. The
MZA correlation for weight does not yield the same estimate as the
ordinary twin method (.51 vs. .80). This suggests that the equal
environment assumption made by the Falconer method has been violated. There is evidence that the assumption is violated for
dietary intake (Fabsitz et al., 1978). All the EEG parameters
except Phi also follow a simple additive model fairly closely.
Lykken (1982) has speculated on why the Phi component (as well as
a number of other traits) does not follow the additive model. The
circadian rhythm parameters (Hanson et al., 1983) suggest substantial hereditary effects, but the correlations are lower than those
for the EEG data. This difference may be due to differences in
reliability of measurement.

Overall, the evidence suggests that identical twins reared
apart are remarkably alike in physical characteristics.

Intelligence

The heritability of human intelligence is one of the most
hotly disputed topics in the behavioral sciences. Jensen (1981)
claims:

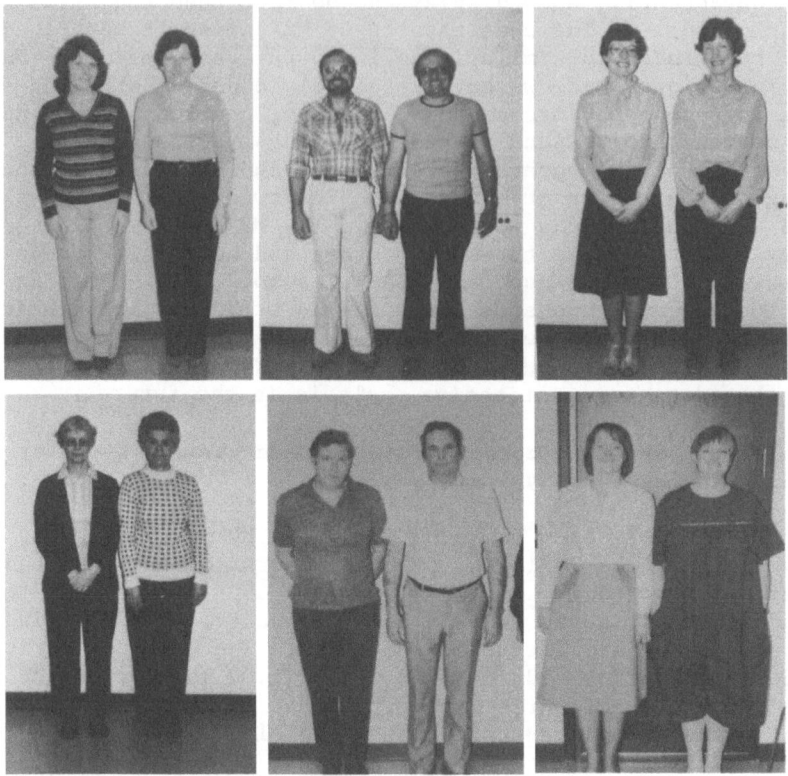

Fig. 3. Three sets of monozygotic twins reared apart (top) and three sets of dyzygotic twins reared apart (bottom). Note the similar posture of the monozygotic twins.

> There is a great deal of agreement among scientists regarding the heritability of intelligence. The experts are not concerned with arguing about any particular estimated value of h^2 within the whole range of most empirical studies; that is, between about .50 and .80. They all recognize the reasons for the variation in estimates of h^2 from one study to another. They are, however, generally in agreement concerning the very substantial heritability of intelligence and IQ. (p. 104)

Jensen, to support his claim, proceeds to cite sixteen different sources. Kamin (Eysenck and Kamin, 1981), on the other hand, claims:

Table 4. Intraclass Correlations for Anthropometric Measures, EEG measures, and Circadian Rhythm Parameters of heart rate for various types of twins

Variables	Correlations DZT	MZT	MZA
Anthropometric Variables	(N=146)	(N=274)	(N=30)
Fingerprint Ridgecount	.46	.96	.98
Height	.50	.93	.94
Weight	.43	.83	.51
EEG Spectrum Parameters	(N=53)	(N=89)	(N=25)
Circadian Rhythm Parameters of Heart Rate			(N=19)
Mesor			.68
Amplitude	No data on ordinary twins		.61
Acrophase			.66

Whatever the "experts" may say, there is no compelling evidence that the heritability of IQ is 80 per cent, or 50 per cent, or 20 per cent. There are not even adequate grounds for dismissing the hypothesis that the heritability of IQ is zero. The evidence is clearly inconsistent with a high heritability. (p. 154)

The principal evidence in favor of a reasonably high heritability comes from twin studies. I will selectively review the relevant evidence and some of the attacks made against it.

Fig. 4 contains a summary of the world literature on familial resemblance in intelligence.

The 34 MZT correlations yield a weighted average of .86. The 41 DZT correlations yield a weighted average of .60. The simple Falconer formula for heritability, $2(r_{MZT} - r_{DZT})$, yields a value of .52. The weighted mean correlation between MZ twins reared apart, however, is .72. Thus there is a sizeable discrepancy between the two estimates. Nevertheless, there is a sizeable genetic influence unless one can show that the studies, or the methodologies, are seriously flawed.

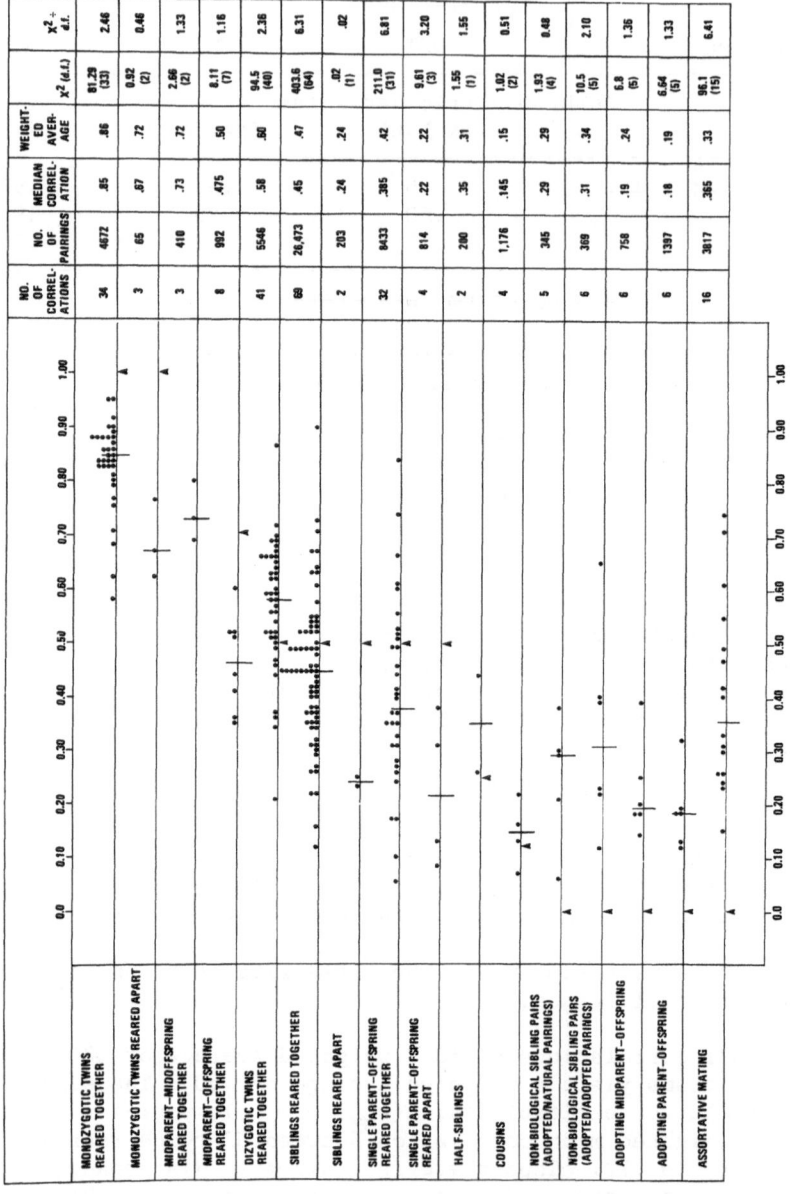

Fig. 4. Familial correlations for IQ. The vertical bar in each distribution indicates the median correlation; the arrow, the correlations predicted by a simple polygenic model. (From Bouchard and McGue, 1981).

TWINS REARED TOGETHER

As discussed in the previous section, the major objection to
the classic twin design is that the environments of MZ twins may
be subtly similar, and therefore a cause of their similarity on
whatever trait is being studied. Such an argument, of course,
should be supported by evidence. What does the evidence look like?

There is no doubt that MZ twins are exposed to more similar
environments than DZ twins (Loehlin and Nichols, 1976; Lytton,
1978, 1980). Kamin (Eysenck and Kamin, 1981), citing Loehlin and
Nichols (1976), points out that,

> Female DZ twins are more likely to sleep in the same bed-
> room than are male DZs. The female DZs are in fact about
> as likely to sleep in the same bedroom as are MZ twins,
> among whom there is no sex difference in the tendency
> to sleep in the same room. (p. 128)

This "suggests to an environmental theorist that the IQ similarity
of female twins should be greater than that of male DZs, even
though DZs of both sexes have about 50 per cent of their genes in
common" (p. 128). To demonstrate his point, Kamin reanalyzed the
data from a study by Huntley (1966) which, coincidentally, did not
report the correlations by sex. The data confirm the hypothesis
(Male DZs, r = .51; Female DZs, r = .70). There are, however, in
the scientific literature (Bouchard and McGue, 1981; Fig. 4 this
chapter) 10 female DZ correlations (730 pairings, median r = .58,
weighted average .61), and 11 male DZ correlations (964 pairings,
median r = .64, weighted average .65). The correlations differ,
but in the wrong direction. The hypothesis is clearly discon-
firmed.

Kamin makes the same prediction for same-sex and opposite sex
twins, with same-sex twins predicted to be more similar. He presents four pairs of correlations, three of which support his
hypothesis (weighted average: same-sex pairs, r = .62, opposite-
sex pairs, r = .47). In fact, there are, however, in the scientific
literature (see Fig. 4 this chapter) 23 published same-sex correlations (3,676 pairings, median r = .61, weighted average = .62)
and 14 published opposite-sex correlations (1,592 pairings, median
r = .565, weighted average = .57). The difference supports the
hypothesis, but not nearly to as great a degree as the figures
reported by Kamin. He now has a 50 per cent hit rate. The same
hypothesis would lead one to expect a similar difference between
same-sex and opposite-sex sibling pairs. There are no differences
whatsoever (weighted average: 6,098 same-sex pairings, r = .48,
5,127 opposite-sex pairings, r = .49). In a large scale twin study
it is quite possible to test systematically for sex differences
in means, correlations and variances and this is the recommended
procedure (Clarke et al., 1980; Eaves, 1982; Martin, 1975).

After summarizing the evidence on this issue, DeFries and Plomin (1978) concluded "Clearly, the burden of proof has shifted to those who claim that the twin method is invalid because identical twins share more similar environments than fraternal twins" (p. 480).

TWINS REARED APART

The most recent reanalysis of the MZA IQ data is by Farber (1981). Her analysis is part of a larger reanalysis of the entire literature on MZAs. I have discussed her treatment of the IQ data at length elsewhere (Bouchard, 1982) and therefore will only treat it briefly here. Farber reports the following conclusions based on her analysis:

> My own evaluation, particularly of the allegedly scientific analyses made of the IQ data, is more caustic. Suffice it to say that it seems that there has been a great deal of action with numbers but not much progress--or sometimes not even much common sense.

In my review of this book, I document in some detail that this conclusion is based on a pseudoanalysis. This can be shown by simply calculating some summary statistics based on Farber's own classifications. She sorted all the available cases into three categories (High, Mixed, and Little) according to degree of separation. She did not, however, report the correlations for these various subgroups. I calculated them and they are shown in Table 5 below.

Table 5. Intraclass correlations for IQ by level of separation
for both sexes and combined group
Classification of separation by Farber

Sex	Degree of Separation		
	Highly	Mixed	Little
Male	.76 (13)	.91 (8)	.84 (9)
Female	.76 (26)	.43 (11)	.65 (16)
Combined	.76 (39)	.43 (19)	.65 (25)

There is no evidence here that the overall correlation of .76 is inflated by reason of lack of separation.

Farber, following Kamin, spends a great deal of time arguing for the invalidity of Shields' measures of IQ (a composite of the Dominoes and Mill-Hill Vocabulary). The effects of dropping Shields data from the highly separated group are shown in Table 6. The inclusion or exclusion of the Shields' data makes no difference whatsoever. The correlation is in fact numerically higher without the Shields' data.

The intraclass correlation, means, and standard deviations for all cases and by sex are given in Table 7. The figure .77 is, in my opinion, the best estimate of the overall estimate of the IQ correlation one might expect in a study of MZA twins. The intraclass correlations for several measures of intelligence in our current study of MZA twins are reported in Table 8 (see Lykken, 1982, for a more detailed discussion of the measures).

Table 6. MZA intraclass correlations, means, and standard deviations for highly separated twins by sex, combined and with Shields' data omitted

Sex	N	r	Mean	S.D.
Males	13	.76	96.35	14.20
Females	26	.76	97.96	14.29
Combined	39	.76	97.42	14.28
Shields' data omitted	28	.78	99.36	14.94

Table 7. Intraclass Correlations for IQ for all MZA twins reported in the literature by sex

Sex	N	r	Mean	S.D.
Males	32	.74	97.6	12.0
Females	51	.79	95.6	14.2
Combined	83	.77	96.3	13.4

Table 8. Intraclass correlations of Raven, Mill-Hill, and derived
scores for three types of twins (see Lykken, 1982, for details)

Intelligence Measure	Correlations		
	DZT (N=42)	MZT (N=71)	MZA (N=29)
Raven	.19	.66	.58
Mill-Hill	.37	.74	.78
IQ(R+M)	.14	.78	.71

Note: The Raven and Mill-Hill are administered as untimed tests.
All scores are corrected for age and sex differences before corre-
lating.

Unlike Kamin, we believe that there is compelling evidence that the
heritability of IQ is well above zero and probably between .50 and
.80, depending on the population being studied.

Personality

The domain of personality is the one in which most psychologists believe that common family environmental factors and social
learning are of great importance in the determination of individual
differences. The following quote is taken from the 1981 edition of
one of the most widely used textbooks in personality:

> Genes and glands are obviously important, but social
> learning also has a dramatic role. Imagine the enor-
> mous differences that would be found in the personali-
> ties of twins with identical genetic endowments if
> they were raised apart in two different families--or,
> even more striking, in two totally different cultures.
> Through social learning, vast differences develop
> among people in their reactions to most of the stimuli
> they face in daily life. As a result of social proc-
> esses, stimuli that terrify one person may delight the
> next and leave a third indifferent. We see, for
> example, an adolescent who seems angered or bored by
> the same goals his parents pursue avidly, and parents,
> in turn, who are horrified by the activities and values
> their daughter seems to treasure. (Mischel, 1981,
> p. 311)

I believe that this point of view is representative and held by
the majority of American psychologists.

TWINS REARED TOGETHER

Nichols (1978) has summarized the world literature dealing with the correlations between twins in the domains of ability, personality, and interests. Table 9 below presents his findings for personality traits.

The simple Falconer heritability for extraversion would be .54, for neuroticism it would be .58. As I indicated earlier, if the raw data are available it is possible to do a somewhat more sophisticated biometric analysis using mean squares. Eaves and Young (1981) have reanalyzed the Extraversion-Introversion and Neuroticism scores of the National Merit Scholarship sample studied by Loehlin and Nichols (1976). They fitted three models to the data. The first model allowed for random within-family effect (E1) and additive gene effects. The second model allowed for both within-family (E1) and between-family (E2 or common) environmental effects, but no genetic effects. The third model specifies the joint effect of all three. The results were quite straightforward. A model with only environmental parameters cannot be fitted to the data for either sex or both sexes taken jointly. The model that assumes no common family environmental effect fits all three cases, and is as capable of fitting both sexes simultaneously (with identical parameter estimates) as it is of fitting the results for the sexes separately. Adding the common environmental effect does not improve the fit of the third model over the second.

Table 9. Mean intraclass correlations for MZ and DZ from twin studies of personality traits

Personality Trait	No. of Studies	r	
		DZ	MZ
Extraversion	30	.25	.52
Neuroticism	23	.22	.51
Socialization	6	.23	.49
Dominance	13	.31	.53
Masculinity-Femininity	7	.17	.43
Hypochondriasis	9	.19	.37
Conformity	5	.20	.41
Flexibility	7	.27	.46
Impulsiveness	6	.29	.48
Median	106	.23	.48

From Nichols, 1978, Table 1.

TWINS REARED APART

There is only a small body of literature dealing with the similarity in personality of MZA twins. Table 10 below summarizes the available data from the Newman et al. (1937) and Shields (1962) studies.

Jinks and Fulker (1970) carried out a biometric analysis of the Shields data and concluded that neuroticism could be fit by a very simple additive model; assortative mating was indicated and common family environment was unimportant. The broad heritability was .54. A simple model could not be fit to the Extraversion data. The negative DZ correlation was interpreted as indicating that the twins were reacting against each other. Common family environment was unimportant. The Newman et al. Neuroticism data are consistent with an interpretation of no common family environmental influence. The three MZA correlations are homogeneous and yield a weighted mean of .57.

In the Minnesota Study the median correlation for twenty-eight MZA twins across eleven Differential Personality Questionnaire scales is .65. Across the same scales the median for DZT (N = a122) and MZT (N = a202) twins are .27 and .53. These results suggest a heritability of about .50 and no common family environmental effect. They replicate the findings with ordinary twins and completely contradict the prediction made by Mischel in the quotation cited above. Can it be true that common family environment has at best only a minor effect on personality?

The answer is yes! Nevertheless, the results from twin studies are not totally consistent with other behavior genetic designs. Scarr et al. (1981a, 1981b) have reported the results of an adoption study that involved biological and adoptive children who were old

Table 10. Intraclass correlations for neuroticism and extraversion for identical and fraternal twins reared apart and together

Twin Type	N	Neuroticism	Extraversion
MZA (Shields)	43	.53	.61
MZA (NF&H)	19	.58	–
MZT (Shields)	43	.38	.42
MZT (NF&H)	50	.56	–
DZA & DZT (Shields)	25	.11	–.17
DZT (NF&H)	50	.37	–

enough to take the same personality tests as their parents. Six
personality scales overlapped with those in our study. The Scarr
et al. data and our MZA and ordinary twin data are shown in
Table 11.

The correlations between genetically unrelated individuals
reflect only environmental influences and suggest a common family
environmental effect of about 5 per cent. The mid-parent X child
and sibling correlations are much larger, but not quite as large as
the DZT correlations. The MZA correlation is, in this instance, the
same size as the MZT correlation. The correlations for first
degree relatives are similar to the summary recently compiled by
Loehlin et al. (1981). They report an unweighted mean correlation
over a number of studies and personality traits of .14 for parent
vs. child and .17 for siblings. Scarr et al. (1981a) argue that
their data suggest a more modest heritability than the twin data.
If we subtract the mean of the adoptive correlations (.05, environmental effect only) from the correlation between first degree
relatives (.17, half the genetic effect plus the environmental
effect), the heritability is .24. The Falconer heritability for
the twin data is .56, 2(21-.49), and is confirmed independently
by the MZA data (.49).

There are a number of possible explanations for these dis-
crepancies. It may be that adoptive parents put far more effort
into influencing the development of personality characteristics
of their children than biological parents (because their children
are naturally less like them?). Consequently, data from adoptive
families would lead us to underestimate genetic influences in the
general population. Biological children and their parents differ
greatly in age. It is quite possible that the same age-corrected
standard score on a personality test has a somewhat different
meaning for respondents of greatly differing ages. This would
attenuate parent-offspring correlations. It would, however, only

Table 11. Comparison of intraclass correlations for selected personality measures from twin and adoption studies

| Scale Name | Adoptive | | Biological | | Twins | | |
	Parent X child	Sibling	Parent X child	Sibling	DZT	MZT	MZA
Median of Six Scales	.04	.07	.15	.20	.21	.49	.49

explain a small part of the difference between the sibling and the
DZ twin correlations.

In their discussion of this problem, Scarr et al. (1982) have
pointed out that,

> If, as the twin studies suggest, the heritability of per-
> sonality traits is 0.5, then the resemblance of genetic
> relatives should be far greater than we found it to be.
> It may be, however, that individual genotypes evoke and
> select different responses from their environments,
> thereby creating genotype X environmental correlations
> of great importance. (p. 118)

Lytton's (1977) work on parental responses to the behavior of their
MZ and DZ twins supports this view. Our experience with MZA twins
is also very compatible with the view that the behavior of the
twins determines the environments which they will experience.

SUMMARY AND CONCLUSIONS

On the basis of the evidence summarized here, as well as a
great deal of other data, it seems reasonable to conclude that the
twin method remains a visible and important tool in the armamen-
tarium of behavioral genetics. Most criticisms of twin methods
have attempted to demonstrate that they lead to overestimation of
the role of heredity as it influences behavioral traits. A close
examination of many of these so-called biases suggests that they
do not necessarily operate in such a manner.

In the domains of physical and psychophysiological traits, the
various twin designs yield consistent and substantial heritabilities,
in the range .65 to .98. These results are not at all surprising.
In the domain of intelligence, we find somewhat less consistent
estimates of genetic influence, but the effects are still quite
strong, somewhere in the range .50 to .80. These results have
engendered a great deal of controversy and criticism, but most of
the criticisms are ad hoc and cannot be constructively replicated.
The Minnesota Study of Twins Reared Apart has, for example, avoided
many of the so-called vitiating methodological flaws of the previous MZA studies (poorly normed tests, testing by the same examiner,
dependence on a single instrument, bias in the recruitment of twins,
etc.); nevertheless, our findings are essentially identical to those
of the previous investigations. In the domain of personality, estimates of heritability are somewhat lower than in the previously
mentioned domains, but still substantial, somewhere in the area of
.50 to .60. These estimates are not entirely consistent with
estimates derived from adoption data. Both the twin studies and
adoption studies do, however, converge on the surprising finding

that common family environmental influences play only a minor role in the determination of personality.

We plan to study many more twins reared apart before we terminate our project so we do not feel compelled to draw any strong conclusions at this time. We have only begun to explore the environmental correlates of our twin differences. We suspect there are some surprises in store in that data. I personally, however, have been impressed both by the great similarity in the identical twins reared apart and by the individuality of each member of a pair. Each twin has turned out to be a unique individual in his or her own right, a full person whose sense of dignity and self-worth has in no way been diminished by the finding of another person with an identical genetic inheritance. Differing experiences have made them unique and different, while identical genes have made them uniquely alike in spite of those experiences.

ACKNOWLEDGMENTS

The Minnesota Study of Twins Reared Apart has been supported by: The University of Minnesota Graduate School; The Spencer Foundation; the Pioneer Fund; and The National Science Foundation (BNS-7926654). The principal investigators are Thomas J. Bouchard, Jr., David T. Lykken, Leonard L. Heston, Elke Eckert, and Duke Tellegen.

REFERENCES

Bouchard, T. J., Jr., and McGue, M., 1981, Familial studies of intelligence: A review. Science 212, 1055-1059.

Bouchard, T. J., Jr., 1982, Review of Farber, S. L., Identical twins reared apart: A reanalysis. Contemp. Psychol. 27, 190-191.

Bouchard, T. J., Jr., 1981, Review of Eysenck, H. J., and Kamin, L., The intelligence controversy. Am. J. Psychol. 95, 346-349.

Burks, B. S., and Roe, A., 1949, Studies of identical twins reared apart. Psychol. Monographs 63, 300.

Clarke, P., Jardine, R., Martin, N. G., Stark, A. E., and Walsh, R. J., 1980, Sex differences in the inheritance of some anthropometric characters in twins. Acta Genet. Med. Gemellologiae 29, 171-192.

Cronbach, L. J., 1957, The two disciplines of scientific psychology. Amer. Psychol. 12, 671-684.

Cronbach, L. J., 1975, Beyond the two disciplines of scientific psychology. Amer. Psychol. 30, 116-1127.

DeFries, J., and Plomin, R., 1978, Behavior genetics. Ann. Rev. Psychol. 29, 473-575.

Eaves, L. J., Last, K., Martin, N. G., and Jinks, J. L., 1977, A progressive approach to non-additivity and genotype-environment covariance in the analysis of human differences. Brit. J. Math Stat. Psychol. 30, 1-42.

Eaves, L. J., 1982, The utility of twins, in: Genetic Basis of the Epilepsies, V. E. Anderson, W. A. Hauser, J. K. Penry, and C. F. Sing, eds. Raven Press, New York.

Eaves, L. J., and Young, P. A., 1981, How stable are personality traits? in: Twin research 3: Part B, Intelligence, Personality and Development, L. Gedda, P. Parisi, and W. E. Nance, eds. R. Liss, New York.

Eysenck, H. J., and Kamin, L., 1981, The Intelligence Controversy. Wiley, New York.

Falconer, D. S., 1960, Introduction to qualitative genetics. Ronald Press, New York.

Fabsitz, R. R., 1978, A twin analysis of dietary intake: Evidence for a need to control for possible environmental differences in MZ and DZ twins. Behavior Genetics 8, 15-25.

Farber, S. L., 1981, Identical Twins Reared Apart: A Reanalysis. Basic Books, New York.

Feldman, M. W., and Lewontin, R. C., 1975, The heritability hang-up. Science 190, 1163-1168.

Galton, F., 1875, The history of twins as a criterion of the relative powers of nature and nurture. Fraser's Magazine 92, 566-576

Ghiselli, E. E., Campbell, J. P., and Zedeck, S., 1981, Measurement Theory for the Behavioral Sciences, Freeman, San Francisco.

Hanson, R. B., Halberg, F., Tuna, N., Bouchard, T. J., Jr., Lykken, D. T., Cornelissen, G., and Heston, L., 1983, Rhythmometry reveals heritability of circadian characteristics of heart rate of human twins reared apart. Ital. J. Cardiol., in press.

Hernshaw, L. S., 1979, Cyril Burt: Psychologist. Cornell University Press, Ithica.

Huntley, R. M. C., 1966, Heritability of Intelligence, in: Genetic and Environmental Factors in Human Ability, J. E. Meade and A. S. Parks, eds. Plenum, New York.

Jensen, A. R., 1980, Bias in Mental Testing. The Free Press, New York.

Jensen, A. R., 1981, Straight Talk About Mental Tests. The Free Press, New York.

Jinks, J. L., and Fulker, D. W., 1970, Comparison of the biometrical genetical, MAVA, and classical approaches to the analysis of human behavior. Psychol. Bull. 75, 311-349.

Juel-Nielsen, N., 1980, Individual and Environment: Monozygotic Twins Reared Apart. International Universities Press, New York.

Kamin, L. J., 1974, The Science and Politics of IQ. Lawrence Erlbaum Associates, Potomac, Maryland.

Loehlin, J. C., Horn, J. M., and Willerman, L., 1981, Personality resemblance in adoptive families. Behavior Genet. 11, 309-330.

Loehlin, J. C., and Nichols, R. C., 1976, Heredity, Environment, and Personality: A Study of 850 Sets of Twins. University of Texas Press, Austin.

Lord, F. M., and Novick, M., 1968, Statistical Theories of Mental Test Scores. Addison Wesley, Reading, Massachusetts.

Lykken, D. T., 1982, Research with twins: The concept of emergenesis. Psychophysiol. 19, 361–373.

Lykken, D. T., Tellegen, A., and Iacono, W. G., 1982, EEG spectra in twins: Evidence for a neglected mechanism of genetic determination. Physiol. Psychol. 10, 60–65.

Lytton, H., 1977, Do parents create, or respond to, differences in twins? Devel. Psychol. 13, 456–459.

Lytton, H., 1978, Genetic analysis of twins' naturalistically observed behavior, in: Twin Research: Part A, Psychology and Methodology, W. E. Nance, ed. Alan R. Liss, New York.

Lytton, H., 1980, Parent-Child Interaction: The Socialization Process Observed in Twin and Singleton Families. Plenum Press, New York.

Martin, N. G., 1975, The inheritance of scholastic abilities in a sample of twins. Ann. Human Genetics 39, 219–229.

McAskie, M., and Clarke, A. M., 1976. Parent-offspring resemblances in intelligence: Theories and evidence. Brit. J. Psychol. 67, 243–273.

Mischel, W., 1981, Introduction to Personality. Holt Rinehart & Winston, New York.

Newman, H. H., Freeman, F. N., and Holzinger, K. J., 1937, Twins: A Study of Heredity & Environment. University of Chicago Press.

Nichols, R. C., 1978, Twin studies of ability, personality and interests. Homo 29, 158–173.

Price, B., 1950, Primary biases in twin studies: A review of prenatal and natal difference-producing factors in monozygotic pairs. Am. J. Human Genetics 2, 293–352.

Plomin, R., DeFries, J. C., and Loehlin, J. C., 1977, Genotype-environment interaction and correlation in the analysis of human behavior. Psychol. Bull. 84, 309–322.

Plomin, R., Willerman, L., and Loehlin, J. C., 1976, Resemblance in appearance and the equal environment assumption in twin studies of personality. Behavior Genetics 34, 43–53.

Roberts, R. C., 1967, Some concepts and methods in quantitative genetics, in: Behavior Genetic Analysis, J. Hirsch, ed. McGraw-Hill, New York.

Rosenthal, D., 1970, Genetic Theory and Abnormal Behavior. McGraw-Hill, New York.

Rowe, D. C., and Plomin, R., 1981, The importance of nonshared (E1) environmental influences in behavioral development. Devel. Psychol. 17, 517–531.

Rose, R. J., 1981, Review of Farber, S. L.: Identical twins reared apart: A reanalysis. Science 215, 959–960.

Rose, R. J., Uchida, I. A., and Christian, J. C., 1981, Placenta-
    tion effects on cognitive resemblance of adult monozygotes.
    in: Twin Research 3: Part B, Intelligence, Personality and
    Development, L. Gedda, P. Parisi, and W. E. Nance, eds.
    Alan R. Liss, New York.
Scarr, S., 1968, Environmental bias in twin studies. Eugenics
    Quart. 15, 34-40.
Scarr, S., 1982, Similarities and differences among siblings, in:
    Sibling Relationships, M. E. Lamb and B. Sutton-Smith, eds.
    Lawrence Earlbaum Associates, Hillsdale, New Jersey.
Scarr, S., and Carter-Saltzman, L., 1979, Twin method: Defense of
    a critical assumption. Behavior Genetics 9, 527-542.
Scarr, S., Webber, P. L., Weinberg, R. A., and Wittig, M. A.,
    1981a, Personality resemblance among adolescents and their
    parents in biologically related and adoptive families.
    J. Personality Social Psychol. 40, 885-898.
Scarr, S., Webber, P. L., Weinberg, R. A., and Wittig, M. A.,
    1981b, Personality resemblance among adolescents and their
    parents in biologically related and adoptive families, in:
    Twin Research 3: Part B, Intelligence, Personality and Devel-
    opment, L. Gedda, P. Parisi, and W. E. Nance, eds. Alan R.
    Liss, New York.
Scarr, S., Scarf, E., and Weinberg, R. A., 1980, Perceived and
    actual similarities in biological and adoptive families:
    Does perceived similarity bias genetic inferences? Behavior
    Genetics 10, 445-458.
Shields, J., 1962, Monozygotic Twins: Brought Up Apart and Brought
    Up Together. Oxford University Press, London.
Shields, J., 1978, MZA twins: Their use and abuse, in: Twin
    Research: Part A, Psychology and Methodology, W. E. Nance, ed.
    Alan R. Liss, New York.
Thomas, H., 1982, IQ, interval scales, and normal distributions.
    Psychol. Bull. 91, 198-202.
Willerman, L., 1979, Effects of families on intellectual development.
    Amer. Psychol. 34, 923-929.

DISCUSSION

STANFORD: I thought your presentation was extremely reassur-
ing.

BOUCHARD: My colleagues in psychology wouldn't say that.

STANFORD: I'm not a psychologist, so I can say that. I have
felt for some time that we've not attributed to genetic influences
their true importance in understanding human behavior; so I feel
myself swimming, well, doing the free style again toward one of
Dan's attractive gradients. Sitting here over the last 24 hours I
have felt reverberating in my mind's ears the ancient cry of the
psalmist. "What is man that thou art mindful of him or the son of
man that thou visitest him." The King James scholars couldn't
bring themselves to translate literally the next line; they couldn't
bring the awesome Hebrew, "Thou hast made him a little lower than
the divine" into literal English, so they translated it as "Thou
hast made him a little lower than the angels," a euphemism. I've
been thinking about the efforts of scholars in Western civilization
over the last two hundred to three hundred years to answer that question; for example, Rousseau's noble savage, Darwin's clever animal,
Marx's economic puppet, Skinner's conditioned responder, Wilson's
genetic straitjacket, Montagu's transmitter of acquired cultural
characteristics. How would you answer the psalmist's query, "What is
man?" or, more particularly, the question implicit in Galton's statement that you quoted, nature or nurture, or what combination?

BOUCHARD: Well, I think our data clearly shows a combination.
I think it's clear that genetic factors are far more important in
the determination of human behavior than we previously suspected,
at least than psychologists had previously suspected. There is
virtually nothing that we look at in these twins that doesn't show
a remarkable degree of similarity. We expected to find some areas
where there was similarity; IQ, for example, has been studied in
some detail. We expected to find some areas where there wouldn't
be any similarity at all, and we haven't found that. (All the
data has not, however, been analyzed.) I think this important conclusion should not overshadow another striking finding in our
study and that is that the members of identical twin pairs, whether
reared apart or together, are unique human beings in their own
right. No matter how similar we found them to be, their fundamental uniqueness has always shown through. Furthermore, for the
twins reared apart, the finding of someone so similar to themselves
has seldom been a problem and has in no way detracted from their
own perception of their uniqueness or their dignity as human beings.
On the personal level, the finding of a co-twin has been a phenome-
nally positive event in their lives. We have some twin pairs who
are strikingly similar but even then they are always distinctive
enough so that while they can enjoy their tremendous similarity

they also sense their own fundamental uniqueness and identity. Some
people have criticized us for emphasizing the tremendous similarity
among the twin pairs. A focus on similarity alone is misleading.
The true picture is much more complicated than that and more satisfying.

STANFORD: That's one of the common denominators of the reli-
gious and political heritage of Western civilization--the significance of individuality.

BOUCHARD: Oh, yes. And, of course, none of the correlations
I've shown you are a "one." None of the findings imply that these
people are so identical that you would confuse them or mix them up.
If you work with these people for five minutes you absolutely know
which one is which. There just isn't any question about it.

LOEWENSTEIN: A neurobiologist accustomed to looking at nerve circuitry, central or peripheral, in lower animals or in higher animals,
would not be too surprised by your findings related to basic instinctive behavior. Based on the little he knows about the genetics of
the basic nerve circuitry underlying reflexes, he would come to
similar conclusions as you have, or at least to complementary ones,
namely, the basic nerve circuitry is genetically determined; and
many of us a priori would have had little hesitation in accepting
that instinctive behavior, by which I mean behavior largely domi-
nated by unconditioned reflexes, is genetically determined.

BOUCHARD: Would you though?

LOEWENSTEIN: I'm referring here only to those really basic
instinctive actions. I'm aware I am using now a vague terminology
of the psychologist.

BOUCHARD: "Instinctive"--that's a tough word; that's a tough
word.

LOEWENSTEIN: As a neurobiologist, I will stick to the term
reflexes. Here the basic neuronal circuitry is--and there is
little question about that--genetically determined. The body of
evidence accumulated over the past thirty, forty years in vertebrates and invertebrates affirm this point. And, indeed, it would
be surprising if the basic circuitry of the nervous system would
be left to the vagaries of the environment. Yet, there is--and
again this is based on hard evidence, as hard as it comes in
biology--there is a certain amount of leeway in the developing
nervous system, a plasticity where the environment is a determinant.
The far-reaching work of David Hubel and Thorsten Wiesel showed
that certain relatively minor deprivations of visual experiences
in the visual system shortly after birth leave profound sequels in

the brain circuitry. When a kitten or a baby monkey were deprived
of pattern vision during the first weeks after birth--by such mild
procedures as sewing half a ping pong ball in front of the eye,
which lets light through to the retina but blurs all images on it--
certain essential connections with cells in the visual cortex never
formed and the animals became forever blind. Lack of image superposition by producing an experimental squint--strabismus is the
medical word--also led to serious permanent visual deprivation.
And there are clinical observations in human babies with squints
that fit this picture, too. So it seems that some of the minor
circuitry, the more plastic part, is influenced by the environment.
And in this sense it may be worthwhile having another look at your
very interesting findings. From your indices and scales--they are
rather complex--it is hard to separate what might reflect basic
and plastic nerve circuitry. I wonder if you could use simpler
indices. I may be asking too much; it is not easy to devise good
and simple behavioral indices. But, perhaps, one could analyze
sequential movement--there are computer techniques available now
for analyzing such things--say movements in a tennis game or simple
movement reactions to noxious stimuli? Would such analysis bring
out differences between siblings or between twins reared apart?

BOUCHARD: I suspect that that may be true, but it is the
case that simple kinds of measures in psychology have always been
shown to be relatively unreliable if they are used to draw inferences about more important traits.

LOEWENSTEIN: Well, I don't know what you call important. I
would say that the simpler measures, the analytically simpler ones,
might be more instructive for the question I discussed, not perhaps what you want. . . .

BOUCHARD: Well, I think you're correct in saying that we
work with largely global characteristics and we're not breaking
behavior down into the fine detail that might show the very specific differences that you're postulating. I think that's
possible.

KOSHLAND: But your EEGs are essentially that, I mean that's
as simple as you can get.

BOUCHARD: I don't think my colleagues in psychophysiology
would say that. They'd say EEGs are concatenations of all kinds
of components and they don't really know how to partition them.

TOBACH: But he's giving you a bundle.

BOUCHARD: Operationally, they're simple.

LOEWENSTEIN: EEG is very complex business.

FOX: Tom, would all that go for Eysenck's work?

BOUCHARD: Eysenck is looking at some evoked potentials. He
correlates some features of the evoked potentials with IQ. We've
just modified all our experimental apparatus to get exactly the
same evoked potential. We have measured evoked potentials on our
previous twins, but apparently Eysenck's procedure is specialized.
I don't think the evoked potentials are necessarily a simple function either. It's not clear what a simple functional measure would
be. For example, reaction time to a tone might appear to be a
simple function but in order to get a reliable measure you have to
run the person through hundreds of trials; it takes an enormous
amount of time. I come to a conference like this and I hear about
how small fluctuations in a couple of molecules within a cell will
have tremendous influence on the way the organism will develop and
then I say, my God! I have this enormously complicated organism
that has gone through a long process of biological development as
well as a complex environment and yet we see some remarkable similarity. There must be stabilizing systems that are driving development into a common channel or something like that. It is not obvious to me how to get from identical genes to 36-year-old individuals
reared apart who are this similar. There's an enormous gap here
that's got to be filled and I think Ethel's going to tell us how
this might be done.

RUTTER: These data put limits, it seems to me, on what stochastic processes affect human beings. There are certainly gross
changes in physiognomy. All of those measurements can occur via,
say, quantal variation which is simply problablistic in genetically
similar individuals. That's really striking. I wanted to pursue
another aspect of the data; that is, if you take nongenetically
related individuals who have, for example, a very closely approxi-
mated EKG or EEG, or some of the other parameters, for example, they
resemble each other physically. Match them on some key aspect and
see if that has any correlation with any of the tests that you made.
It seems to me that that's an important control for the kind of
thing that you count.

BOUCHARD: Most of the data we talk about and specifically
the personality and IQ data, is age and sex corrected. For example,
if you administer ability tests, particularly to children, you find
the older children do better than the younger children. Simply
because twins are matched in age, they are going to be similar in
ability. So you have control for age.

RUTTER: I want to pursue this for a minute. Can you actually find a set of individuals with essentially identical profiles?
I doubt if you would find identical scores but hopefully you would
find some that are very similar.

TOBACH: We have got to point out a position that I think
we do not disagree on, as found in personality and psychological
work all the time. There are clusters of tests that go together
and there are profiles of people who are not related to each other
in which you can find relationships among certain measures. So, in
a sense, there are core patterns.

BOUCHARD: They call it types; I'm not a typologist.

TOBACH: I am not a typologist either, but the point is
that the approach that has been used traditionally over and over
again finds that kind of clustering. Now nobody has, as far as I
know, any correlation with any of the lower-level biochemical, biophysical types. In other words, when you go into things like
reaction-time experiments, temperature-control reaction, you don't
get such a good correlation with these clusters.

BOUCHARD: I think you were saying: "Match them up on EEG,
see how they look on personality, how they look on interest. Is
that what you were asking? That hasn't been done as far as I know.
I would be willing to lay a lot of money right now that it will not
work.

HARDIN: You should match them up on a "blind" test, and
we should run through a computer program.

KOSHLAND: First, I think it is a terrific study. I don't
quite agree with Werner. I think that maybe Werner was way ahead
of his time and would have known this, but my own dealing with
psychologists indicates they resist an intelligence-heredity correlation heavily even though rats can be put through mazes and
selected for the brighter rats.

TOBACH: It's very difficult to talk for psychologists.

KOSHLAND: Second, I think it means that you can get absolute
determinism if you have a large enough number of molecules. If
the process has been controlled by 10^{10} enzyme molecules, the statistical fluctuation is very small. And so, it seems to me that
your evidence now is suggesting a Poissonian variation in the individuality of twins because you almost eliminate environment. With
twins reared apart, the correlation coefficient is 0.5 to 0.7, but
the correlation with environment is .05. Maybe there in the individuality is some Poissonian variation in some of these personality
traits. Thus, twins reared apart will have this heavy core of
properties which are similar and then divergences which are not due
to the environment. They are due to something innate in the twins.

BOUCHARD: That's within the realm of possibility, no question about it. I would tend to look more toward prenatal factors

as an important source of differences. It would be a hard hypothesis to prove or disprove.

KOSHLAND: To follow what Bill Rutter was saying, I looked at
your correlation on achievement and it didn't seem to be any better
than chance. Achievement is the kind of thing that middle-class
parents pushing kids might influence, so it would correlate with
environment, not heredity.

BOUCHARD: We need more cases before we can do an analysis
of differences in heritability.

ENDOGENOUSLY DETERMINED VARIANTS AS PRECURSORS

OF SUBSTRATES FOR NATURAL SELECTION

Sidney W. Fox and Tadayoshi Nakashima

Institute for Molecular and Cellular Evolution
University of Miami
Coral Gables, Florida 33134

INTRODUCTION

The emergence of living individuals would seem to have been synonymous with what is usually referred to as the origin of life. Experiments in the formation of protoorganisms suggest, however, that individuality can be identified for an even earlier molecular stage in immediate precursors of those first organisms. Those precursors, which avidly assemble to "laboratory protocells" in the presence of water, are known as proteinoids, or are indexed in Chemical Abstracts as thermal proteins.

The only relatively comprehensive laboratory model in existence for the emergence of cellular life has as its first steps the formation of thermal proteins by geological warming of any of a wide variety of suitable mixtures of amino acids (Florkin, 1975; Lehninger, 1975; Fox and Dose, 1977; Fox, 1978; Follmann, 1982; Fox, 1983). These thermal proteins have been found to be highly ordered* within molecules and between molecules. They also have arrays of enzymelike activities (Dose, 1984) and they easily form structures suggesting protocells (Fox and Dose, 1977), which are in some cases aggregations of peptides capable themselves of synthesizing peptides, plus polynucleotides (Fox et al., 1982). The emergence of a genetic mechanism and code is thereby suggested (Fox, 1981).

---

*In this paper, the terms self-limiting and internally directing are taken as synonymous; they are a significant aspect of intramolecular and intermolecular ordering.

The thermal proteins are the only candidates we yet have, on the basis of experiments, for the first biologically informative macromolecules ab initio. In this paper, we will review the range of variability in some of the thermal proteins made in the laboratory. The results all indicate that the products are nonrandom (Melius, 1979), i.e., they are self-limited in type. This fact, in association with the results in further conversion of thermal protein, offers the possibility of explaining a primary contribution to the emergence of individuality as we know it.

EMERGENCE AND EVOLUTION OF LIFE

The data to be reviewed can best be understood in the light of the proteinoid theory that has developed from the production and characterization in the laboratory of thermal proteins. The proteinoid theory has in turn emerged operationally from attempts to retrace the steps of molecular change involved in the "origin of life." The theory has explained how thermal reactions of amino acid mixtures have yielded proteinlike polymers which readily aggregate to form microstructures looking like cocci (Winchester, 1977), having abilities to reproduce by heterotrophic feeding (Fox et al., 1967, 1974), to manufacture as stated both peptides and polynucleotides (Fox et al., 1982), to display excitability (Przybylski et al., 1982), and to exhibit numerous other properties once thought to be unique to the modern biological unit. Again, the concept of central relevance is that of nonrandomness, in contrast to the randomness assumed within Neo-Darwinian (Fox, 1984).

As has often been pointed out, totally invariant reproduction would not provide the array necessary for natural selection to operate; no evolution could have resulted (Fox, 1953). Invariant reproduction is the equivalent of totally determined types, and is also the equivalent of complete nonrandomness.

THE CHICKEN-EGG QUESTION

Of various evolutionary questions answered by the self-ordering of amino acids (Fox, 1978), the biochemist's chicken-egg question had remained unanswered for a long time. In the modern cell, proteins need nucleic acids whereas nucleic acids need proteins. The two needs are not equivalent, however. The modern requirement for nucleic acids is to specify, or help specify, the sequence of amino acids in protein. On the other hand, the protein is needed to catalyze reactions, including the synthesis of nucleic acids.

Two possibilities for resolution of this dilemma could be seen. For one of them, catalytic activity in nucleic acids has been postulated. Demonstrations of catalytic activity in polyguanylic acid (Mizutani and Ponnamperuma, 1978) and a qualified catalytic activity in RNA (Kruger et al., 1982) have been reported. Each of these activities is, however, sharply limited; they do not truly begin to explain how RNA might be synthesized along a DNA or RNA template. The first example indicates an enhancement in the making of peptide bonds, the second is not useful for that nor for RNA synthesis. In the words of the discoverer of the second activity, Cech (Lewin, 1982), "It's a big jump . . . to an RNA molecule marching down another RNA making a copy of it." Also, for some, the RNA activity is more metathetical than catalytic. The development of true enzymic catalysis in evolution was for hundreds of reactions; the only class of compound now known to be capable of catalyzing such an extensive array is the proteins.

While we thus infer manifold protein enzymes to catalyze manifold specific metabolic reactions, including RNA synthesis, we do not know how many reactions were necessary for a protometabolic organism. We can be quite sure that whichever route of evolution happened to have a class of catalyst capable of wide versatility would have had a selective advantage in evolution. Accepting that this versatility is now in the protein class, one is left with the need to explain ordering processes that would have originally yielded specific repeated peptide chains, comparable to what the modern nucleic acid mechanism achieves. In this modern mechanism, the nucleic acid sequence is decoded, or translated, into the amino acid sequence. The instructions are lodged in the nucleic acids, or rather in the nucleic acids providing instructions through the specificity-conditioned (Kornberg, 1980) protein machinery of the cell (Commoner, 1965). In responding to Commoner's criticism of a narrow emphasis on DNA, Crick (1970) has disavowed that the Central Dogma applies to the origin of life or to the origin of the code.

For the chicken-egg question of original protein or original DNA first, the answer was first suggested in 1956 (Fox), and in incomplete form in 1953 (Fox et al.). The simple, experimental answer is that the first proteins obtained their instructions from the amino acids themselves. This process would necessarily have been an endogenous one, and therefore more appropriate to evolution than a process dependent upon exogenous nucleic acid (assuming that a protonucleic acid could be shown to be helpful in a primitive catalytic situation). What has resolved the impasse is the repeatedly demonstrated self-sequencing of amino acids (Fox, 1981). Evidence for that phenomenon and for self-ordering of amino acids (ordered results without studies of sequence) are reviewed here. The accumulated results permit developing a sense of quantitative degrees of variation. The assessment of variation, i.e., degree

of departure from total internal fixing of sequence will be discussed also as a first molecular evaluation of contributions of
deterministic processes.

SELF-SEQUENCING

    Self-sequencing is a type of self-ordering and is an early
step in the processes involved in self-organization. We distinguish
self-sequencing from self-ordering by the nature of the evidence.
In self-sequencing, the data consist partly or entirely of the
assessment of the positions of individual amino acid types in the
peptide chains. In self-ordering, other types of evidence are used,
e.g., limitation in heterogeneity of polymers. However, if the
amino acid sequences were not ordered, the polymers would theoretically appear as disperse upon fractionation.

    The most thorough study of randomness of sequence in thermal
copolymerization of amino acids is that of Nakashima et al. (1977).
In this work, the amino acids glutamic acid, glycine, and tyrosine
were thermally copolymerized. The peptide mixture obtained was
distributed on a paper chromatogram (Fig. 1). The dominant fraction was tripeptide; no other tripeptide fraction was present. The
tripeptide proved to be an equimolar mixture of two tripeptides.
The results expected and those found are in Table 1. The results
found are extremely nonrandom.

    These results were confirmed by Hartmann et al. (1981). Others
in the same group found that the pigment that is always formed when
amino acids are jointly heated consists of flavin and pterin evi-
dently built into the chain (Heinz et al., 1979). Other evidence
to be reviewed will show that the self-ordering applies throughout
long chains. The Hartmann results, however, suggested that the
tripeptides isolated could be contained within pigment nodes in
longer polymers.

    The weight of nonrandom peptide found relative to that calcu-
lated on the basis of assumed randomness is 19.2 (Nakashima et al.,
1977).

    Another kind of evidence for self-sequencing is that in which
the N-terminal residue type is compared quantitatively with the
C-terminal residue type, or each or both of these are compared
with the total amino acid composition. A number of studies on nonrandomness in whole proteinoids, as established by terminal amino
acid analyses, have been reported (Fox and Harada, 1958; Harada
and Fox, 1960; Fox and Harada, 1960a,6; Fox, 1974; Harada and Fox,
1975; Fox and Suzuki, 1976; Fox et al., 1978).

    All of them indicate nonrandomness (Melius, 1979, 1982).

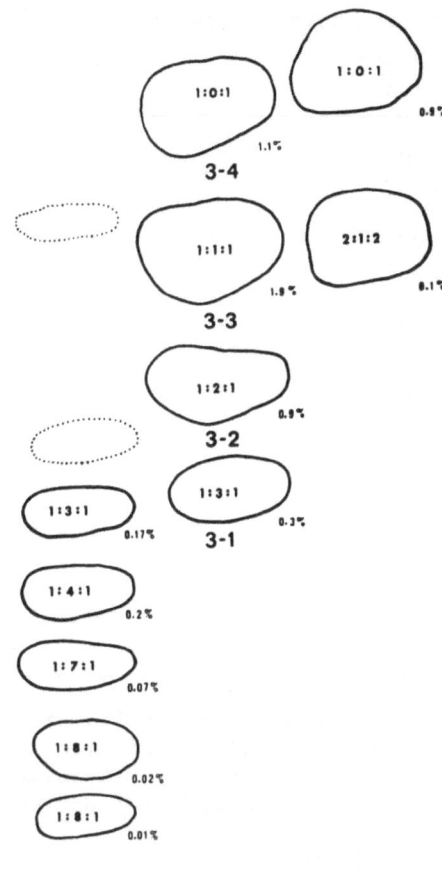

Fig. 1. Discrete peptides by paper chromatography of thermal product of glutamic acid, glycine, and tyrosine (Nakashima et al., 1977; Hartmann et al., 1981). Dominant fraction is 3-3; it represents all tripeptides formed.

However, the above analyses were performed upon unfractionated proteinoids. The studies of Melius and Sheng were performed on fractionated proteinoids and were more definitive, expecially for the context of individuality. Melius and Sheng copolymerized six amino acids: alanine, glutamic acid, glycine, leucine, proline, and phenylalanine. In the product they obtained by fractionation on paper chromatograms, only three peptides (Fig. 2) appeared.

These three peptides were all above 5,000 in molecular weight (Table 2). The peptides were discrete and, in contrast to results

Table 1. Tyrosine-containing tripeptides found vs. those expected
          on the basis of the random hypothesis

Expected from Random Polymerization		Found from Nonrandom Polymerization
αUαUY	YαUU	
αUγUY	YγUU	
γUαUY	YαUG	
γUγUY	YγUG	
αUGY	YαUY	
γUGY	YγUY	
αUYU	PαUY	
γUYU	PγUY	
αUYG	PGY	PGY
γUYG	PYU	
αUYY	PYG	PYG
γUYY	PYY	
GαUY	YGU	
GγUY	YGG	
GGY	YGY	
GYU	YYU	
GYG	YYG	
GYY	YYY	

Note: The dominant fraction obtained from the thermal copolymerization of glutamic acid, glycine, and tyrosine proved to be an
equimolar complex of pyroglutamylglycyltyrosine and pyroglutamyl-
tyrosylglycine (Nakashima et al., 1977; Hartmann et al., 1981).

        U = glutamic acid residue
        Y = tyrosine residue
        G = glycine residue
        P = N-pyroglutamyl

from unfractionated peptides, each had but a single amino acid in
each terminal residue. In the N-terminal position, the only amino
acid was pyroglutamic acid. Three individual peptides could, of
course, have no more than three C-termini in total. Three of the
six conceptual possibilities were thus found. Each C-terminus was
singular: glycine, alanine, or leucine. Since multiple C-terminal
types are found analytically in unfractionated proteinoids (Harada
and Fox, 1975), the singularity suggests three single peptides.
Proline and phenylalanine were totally absent from the C-terminal
and N-terminal positions. Glutamic acid, which in the pyro form
is known as an N->C polymerization initiator (Fox et al., 1978),
is totally absent from all three C-termini.

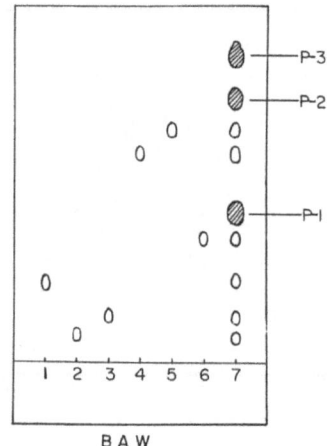

BAW

Fig. 2. One-dimensional paper chromatography of Melius-Sheng peptides. Chromatography of the peptides in BAW (4:1:5). Open spots, ninhydrin positive; shaded spots, ninhydrin negative, HClO-EtOH-starch-K1 positive. 1, Ala; 2, Glu; 3, Gly; 4, Phe; 5, Leu; 6, Pro; 7, thermal polypeptides P-1, P-2, and P-3.

Table 2. Three polypeptides obtained by heating of six amino acids

Polypeptide	Mol. Wt.	N-Terminus	C-Terminus
P-1	5,200	(pyro) Glu	Gly
P-2	10,000	(pyro) Glu	Ala
P-3	11,500	(pyro) Glu	Leu

The remarkable degree of self-sequencing indicated by these experiments is corroborated by the results with copoly(glu, gly, tyr) of Nakashima et al. (1977). It is corroborated in another way in another experiment in which it was found that such phenomena are expressed for basic peptides and acidic peptides separately from a single mixture of amino acids (Snyder and Fox, 1975).

SELF-ORDERING

Much evidence on self-ordering was accumulated prior to the studies of self-sequencing (Dose, 1984). An example is presented in Fig. 3, in which an amidated proteinoid is distributed on DEAE-cellulose. A repetition of the distribution is seen in the dotted

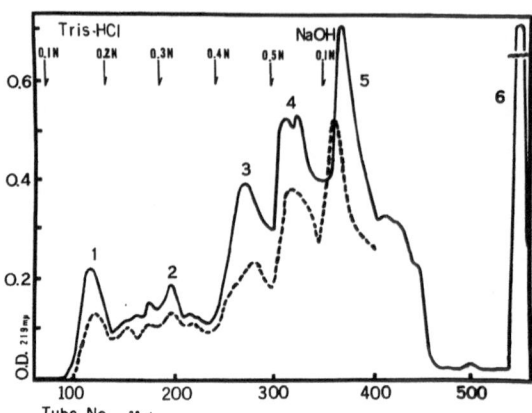

Fig. 3. Fractionation of amidated 1:1:1-proteinoid on DEAE-cellulose column. A few major fractionations are observed. Repetitions
yield similar patterns. The dotted line represents one of these.

graph (Fig. 3). This represents a high degree of reproducibility,
as do all other assessments of repeated synthesis of specific proteinoids. Such results are in strong contrast to the lack of
reproducibility that Eigen (1971), for example, assumes. Of
course, if one assumes randomness, it follows that the synthesis
has no internal directedness. The two related assumptions of randomness and nonreproducibility are exactly what is not, and has
not been, supported by experimental results such as this one.*

    Fig. 4 is a distribution of turtle serum proteins from a
DEAE-cellulose column (Block and Keller, 1960). The mirrored
similarity to Fig. 3 is notable. The specificities in modern
protein synthesis mechanisms, accordingly, may not be greater than
in these model primordial systems.

    Even when the individual peaks 3, 4, and 5 of Fig. 3 are
analyzed, they prove to be very similar to each other. In Fig. 5,
peptide maps of the partially hydrolyzed material eluted from
peaks 3, 4, and 5, are displayed. Comparable similarities were found
in amino acid composition and electrophoretic analyses (Fox and
Nàkashima, 1967) between peaks 3, 4, 5, and the unfractionated
polymer. It thus follows that repetitions of syntheses give
similar collections of major fractions and that, furthermore, the

_____

*The specificities are observed by analytical characterization of
the static products of the processes; by inference, the processes
are dynamically specific.

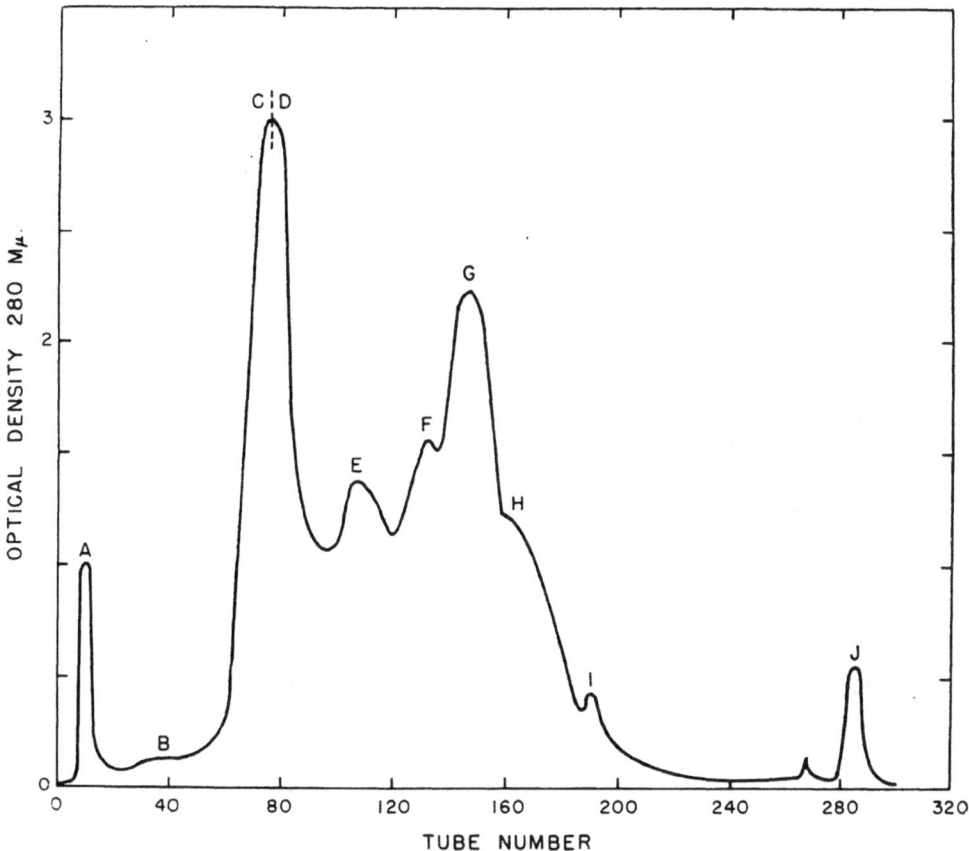

Fig. 4. Distribution of turtle serum proteins on DEAE-SF-cellulose column in sodium phosphate buffer (Block and Keller, 1960).

variation between the individual components obtained in any fractionation is very small. The forces of self-limitation are powerful; they yield limited arrays of individuals which differ only slightly from each other.

MOLECULAR SELECTION AND NATURAL SELECTION

The theory of organismal evolution according to natural selection has been interpreted over a wide range. This catholicity of interpretation was true during the lifetime of its prime formulator, Charles Darwin.

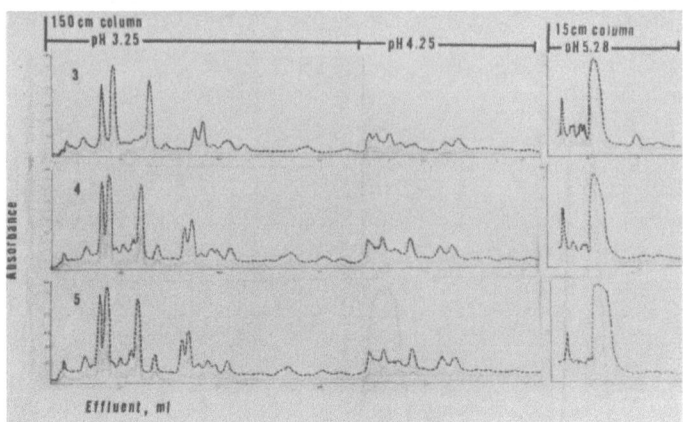

Fig. 5. Peptide map of partial hydrolyzates of peaks, 3, 4, and 5 from eluate depicted in Fig. 3.

The original title of Darwin's famous book is The Origin of Species by Means of Natural Selection. It is known by this title or simply by The Origin of Species or Even The Origin. But the alternative title to be found in the Random House version (Darwin, undated) has the subtitle of: "The Preservation of Favored Races in the Struggle for Life."

On page 52, Darwin refers to preservation by the term natural selection "in order to mark its relation to man's power of selection." But the emphasis is on preservation, a process which is not man-directed, as the word selection unfortunately suggests.

And on page 122, Darwin recognized the basic incompleteness of representing evolutionary thought by natural selection with the words, "Our ignorance of the laws of variation is profound."

A number of theorists have interpreted natural selection as a causative, or creative, agent in evolution (cf. Ho and Saunders, 1979). Others have seen it as a passive sieve (Morgan, 1932; cf. Dobzhansky, 1981). Most or all of the treatments that visualize natural selection as creative also regard the matrix and variation as random. If variants were random, natural selection is necessarily the only process in purview that could be the source of variation, or directedness, as Dobzhansky (1981) referred to it.

The experiments, however, show that the variants are nonrandom; they have directedness (Fox, 1983). The variants at the first stage can be said to have resulted from molecular selection (Fox and Dose, 1972). Molecular selection is in itself nonreproductive and is thereby distinguished from natural selection.

Darwin (undated, p. 372) was in his time unable to explain variation, as he said in speaking of the time when natural selection would be accepted, "A grand and almost untrodden field of inquiry will be opened, on the causes and laws of variation." Morgan (1932, p. 140) was unwilling to regard natural selection as a causative agent. Having consolidated the theory of the chromosomal gene, he said: "Neither the genetic factors responsible for a part of the initial variability, nor the environmental factors, can bring about such an advance." And again (Morgan, 1932, p. 140), "There is here no implication that natural selection itself is responsible for the appearance of new types," and (p. 127), "But only by a perverse use of the term could natural selection be interpreted to mean an active agent in nature."

The experimental fact of self-ordering of amino acids to yield limited numbers of types of thermal proteins having specific catalytic activities permits an updating of Darwin's concept. The variation is in the synthesis and assembly; natural selection is then separately active upon the variants. The variants most adapted to their environment are preserved ("selected") and participate in their reproduction (Commoner, 1965).

The endogenously produced variants in synthesis and assembly are thus the substrates for natural selection. Endogenous variation, in this view, is a full and primary partner in a two-stage evolutionary process.

DETERMINISM

The concept of determinism is easily acceptable to many physical scientists for the level at which they work. One of the early analysts of determinism was Erwin Schrödinger (1946) whose book on What Is Life? included an epilogue on determinism and free will. These are usually regarded as complementary areas, but they need not be so (Matsuno, 1984).

A kind of physical scientist other than the physicist Schrödinger are the biochemists who concern themselves with the unfolding of evolution along a molecular line. It is signal that Kenyon and Steinman (1969), in producing a book on how life emerged, chose to title it Biochemical Predestination.

Another expression of this view is the aphorism occasionally heard among chemists in the NASA space program, "Chemistry is chemistry wherever and whenever it occurs." Organic chemists, especially, who see high reproducibility in the products of the components they put together in their flasks or vats, tend toward molecular determinism although few of them have concerned themselves with this principle.

The operation of determinism, like the operation of evolution, can be seen most clearly at the molecular level, where it is subject to uncomplicated experiments. The connection of molecular determinism to genetic determinism has been much less clear. A modern proponent of genetic determinism, E. O. Wilson (1978), perhaps could not see the connection so clearly because he roots his genetic determinism in DNA (Wilson, 1975), for which an evolutionary origin had not been suggested on the basis of experiments until 1981 (Fox). In his dependence upon DNA, Wilson necessarily eschewed original nonrandomness as had Eigen (1971) and Monod (1971), who saw DNA as the means for overcoming assumed randomness. In skirting nonrandomness (Fox, 1978; Matsuno, 1982), Wilson thus necessarily eschewed molecular determinism. The original genetic determinist, T. H. Morgan (1932, pp. 59, 112, 134), did not rely on randomness, so that Morgan did make the connection as fully as he could have in his time.

The different emphases of Morgan and of Wilson raise a question of whether this can be the reason that Wilson does not mention Morgan in his writing on determinism.

More evidence for nonrandomness between the stages of proteinoid matrix and of nonrandom mutation (Morgan, 1932) is yet wanted, but the establishment of nonrandomness (= determinism) at these two stages is in contrast to the totally unsupported assumptions of total randomness by Eigen, Crick, and others.

If one accepts the connection between molecular determinism, itself most strongly established, and genetic determinism and then recognizes the connection between genetic determinism and behavioral determinism, he is able to perceive a daisy-chain of determinism that proceeds from molecules → organisms (Koshland, 1984) → genes (Morgan) → behavior (Bouchard, 1984). The roots of human behavior are thus in the molecules. The specificities are in the shapes of molecules and in their interactions (Fox, 1983).*

We may ask if all along this daisy-chain of determinism the relationships are comparable to those in molecular determinism. If they are, these state that the result is predetermined, or largely predetermined by its matrix (Fox, 1980b). Not only the result, however, but the distribution in the variability that that result provides (e.g., Fig. 3) is also largely predetermined by its matrix (and to a minor degree, by exogenous conditions).

Arguments for the opposing point of view, against biological determinism, have been assembled in a volume titled Against Biological Determinism (Rose, 1982). The volume seeks to answer

---

*See footnote 2.

what is referred to as a growing body of deterministic arguments within biology, medicine, and sociology. It seems to argue that the form of society of which the individual is part has large effects on individual behavior. The determinist, however, can counterargue that the behaviors of individuals in a society are the expression of the <u>interactions</u> of individual behavioral determinisms; the results of interactions constitute a higher order of complexity. The societal complex of individual determinisms is, then, much more difficult to analyze than any behavior at a lower hierarchical level, but the incomplete analysis does not necessarily invalidate the perceived individual determinisms. We must still respect the fact that, at the molecular level (and at higher levels by extrapolation), the reactions are quite fully reproducible, and therefore molecularly determined.

The deterministic point of view represented by Wilson was in its way also necessarily incomplete since it relied on DNA of unexplained origin and indirectly on the unsupported notion of randomness. The newer view referred to here as daisy-chain determinism purports that determinism, like evolution (Fox, 1981a) is intrinsically hierarchical. That is, the principles of predetermined action continually manifest themselves, but in differing ways at ascending levels of the hierarchy.

The initial directedness, in any event, is rooted in the shapes of the monomeric molecules (Fox, 1980a, 1983) and is developed as manifestations of the endogenously controlled interactions.

None of this is to argue that evolution would proceed on the identical line throughout in any repetition of evolution on this Earth, could such a series of events recur. Within this volume, Koshland (1984) has inferred, from his experimental studies of bacteria especially, the modifying influences of conditions in the environment, and the effect of chance factors on a dominantly deterministic heredity. These results seem to indicate that endogenous programming is a more powerful determinant than scientists in general, or the public, have been accustomed to entertain in their thinking. The results tend to surprise us, especially when they appear in the remarkable behavioral similarities of reunited homozygous twins (Bouchard, 1984).

Scientists operating at other levels (Schrödinger, 1946; Dobzhansky, 1981) have argued for determinism, and against a total determinism. The fact that we can now evaluate determinism at various hierarchical levels, and in some cases begin to relate the quantitation from one level to another, offers us new hope in the investigation invited by the growing understanding of the relationship of individuality to its endogenous programming.

We can also entertain the possibility that organismal evolu-

tion is properly moved theoretically from the gradualism that
Darwin imputed to it. Natural selection appears, accordingly, to
operate on nonrandom variants (Fox, 1983), as well, perhaps, or
whatever chance variants may arise and these variants may represent
either macro or micro modifications of their ancestors.

PHILOSOPHICAL INFERENCE

     This conference at first concentrated upon the chemical and
biological bases of individuality. Individualism and the variety
of individualities are manifestations of liberty at the human
level. If it be true, as some of us believe, that limiting forces,
i.e. nonrandomness and determinism, have operated in evolution, we
can begin to perceive the existence of limitations to the scope of
individualities. Liberty, in an evolutionary sense, then is like-
wise constrained by intrinsic and inherent limits. As this conference developed, the papers and discussions increasingly recognized the possibility of a deterministic evolutionary sequence on
this planet. The doctrine of determinism would thus reflect both
the flowering of individuality and constraints for that flowering.
In the context of this broadened thinking, this book came to be
titled INDIVIDUALITY AND DETERMINISM instead of INDIVIDUALITY.

     We can now perceive the possibility of quantitatively assess-
ing the boundary conditions of individuality and of determinism.
These, and related studies, are tasks for the future.

ACKNOWLEDGMENTS

     The experimental work was made possible by Grant No. NGR
10-007-008 of the National Aeronautics and Space Administration,
by the National Foundation for Cancer Research, and by Mr. David
Rose. Contribution No. 359 of the Institute for Molecular and
Cellular Evolution. Thanks for critical review are expressed to
Professor Paul Melius of Auburn University.

REFERENCES

Block, R. J., and Keller, S., 1960, Interrelations between serum
     protein fractions isolated by various techniques. Evidence
     for amino acid anlagen. Contribs. Boyce Thompson Inst. 20,
     385-394.
Bouchard, T. J., Jr., 1984, Twins reared together and apart: What
     they tell us about human diversity, in: Individuality and
     Determinism, S. W. Fox, ed., Plenum Press, New York.

Commoner, B., 1968, Failure of the Watson-Crick theory as a chemical explanation of inheritance. Nature 220, 334-340.

Crick, F., 1970, Central dogma of molecular biology. Nature 227, 561-563.

Darwin, C., undated, The Origin of Species by Means of Natural Selection or The Preservation of Favored Races in the Struggle for Life. Random House, New York.

Dobzhansky, T., 1981, From potentiality to realization in evolution. Anthropology UCLA 7, 71-82.

Dose, K., 1984, The evolution of individuality at the molecular and protocellular levels, in: Individuality and Determinism, S. W. Fox, ed., Plenum Press, New York.

Eigen, M., 1971, Molecular self-organization and the early stages of evolution. Quart. Rev. Biophys. 4, 149-212.

Florkin, M., 1975, Ideas and experiments in the field of prebiological chemical evolution, in: Comprehensive Biochemistry, Vol. 29, Part B, pp. 231-260, M. Florkin and E. H. Stotz, eds., Elsevier, Amsterdam.

Follmann, H., 1982, Deoxyribonucleotide synthesis and the emergence of DNA in molecular evolution. Naturwissenschaften 69, 75-81.

Fox, S. W., 1956, Evolution of protein molecules and thermal synthesis of biochemical substances. Amer. Scientist 44, 347-359.

Fox, S. W., 1974, Origins of biological information and the genetic code. Mol. Cell. Biochem. 3, 129-142.

Fox, S. W., 1978, The origin and nature of protolife, in: The Nature of Life, pp. 23-92, W. H. Heidcamp, ed., University Park Press, Baltimore.

Fox, S. W., 1980a, Metabolic microspheres. Naturwissenschaften 67, 378-383.

Fox, S. W., 1980b, Life from an orderly Cosmos. Naturwissenschaften 67, 576-581.

Fox, S. W., 1981a, Copolyamino acid fractionation and protobiochemistry. J. Chromatogr. 215, 115-120.

Fox, S. W., 1981b, From inanimate matter to living systems. Amer. Biol. Teacher 43(3), 127-135, 140.

Fox, S. W., 1981c, Origins of the protein synthesis cycle. Int. J. Quantum Chem. QBS8, 441-454.

Fox, S. W., 1984, Creationism and evolutionary protobiogenesis, in: Science and Creationism, A. Montagu, ed., Oxford University Press.

Fox, S. W., 1984, Proteinoid experiments and evolutionary theory, in: Beyond Neo-Darwinism, M-W. Ho and P.T. Saunders, eds., Academic Press, London.

Fox, S. W., and Dose, K., 1972, Molecular Evolution and the Origin of Life, Freeman, San Francisco.

Fox, S. W., and Dose, K., 1977, Molecular Evolution and the Origin of Life, rev. ed., Marcel Dekker, New York.

Fox, S. W., and Harada, K., 1958, Thermal copolymerization of amino acids to a product resembling protein. Science 128, 1214.

Fox, S. W., and Harada, K., 1960a, The thermal copolymerization of amino acids common to protein. J. Amer. Chem. Soc. 82, 3745-3751.

Fox, S. W., and Harada, K., 1960b, Thermal copolymerization of amino acids in presence of phosphoric acid. Arch. Biochem. Biophys. 86, 281-285.

Fox, S. W., and Nakashima, T., 1967, Fractionation and characterization of an amidated thermal 1:1:1-proteinoid. Biochim. Biophys. Acta 140, 155-167.

Fox, S. W., and Suzuki, F., 1976, Linkages in thermal copolymers of lysine. BioSystems 8, 40-44.

Fox, S. W., Winitz, M., and Pettinga, C. W., 1953, Enzymic synthesis of peptide bonds. VI. The influence of residue type on papain-catalyzed reactions of some benzoylamino acids with some amino acid anilides. J. Amer. Chem. Soc. 75, 5539-5542.

Fox, S. W., McCauley, R. J., and Wood, A., 1967, A model of primitive heterotrophic proliferation. Comp. Biochem. Physiol. 20, 773-778.

Fox, S. W., Melius, P., and Nakashima, T., 1978, N-Terminal pyroglutamyl residues in proteins and thermal peptides, in: Evolution of Protein Molecules, pp. 111-120, H. Matsubara and T. Yamanaka, eds., Japan Sc. Soc. Press, Tokyo.

Harada, K., and Fox, S. W., 1960, The thermal copolymerization of aspartic acid and glutamic acid. Arch. Biochem. Biophys. 86, 274-280.

Harada, K., and Fox, S. W., 1975, Characterization of functional groups of acidic thermal polymers of α-amino acids. BioSystems 7, 213-221.

Hartmann, M., Brand, M. C., and Dose, K., 1981, Formation of specific amino acid sequences during thermal polymerization of amino acids. BioSystems 13, 141-147.

Heinz, B., Ried, W., and Dose, K., 1979, Thermal generation of pteridines and flavins from amino acid mixtures. Angew. Chem. Intl. Ed. Engl. 18, 478-483.

Ho, M. W., and Saunders, P. T., 1979, Beyond Neo-Darwinism: An epigenetic approach to evolution. J. Theor. Biol. 78. 573-591.

Kenyon, D. H., and Steinman, G., 1969, Biochemical Predestination. McGraw-Hill, New York.

Kornberg, A., 1980, DNA Replication. Freeman, San Francisco, 87-88.

Koshland, D., Jr., 1984, Individuality in bacteria and its relationship to higher species, in: Individuality and Determinism, S. W. Fox, ed., Plenum Press, New York.

Kruger, K., Grabowski, P. J., Zang, A. J., Sands, J., Gottschling, D. E., and Cech, T. R., 1982, Self-splicing RNA: Autoexcision and autocyclization of the ribosomal RNA intervening sequence of Tetrahymena. Cell 31, 147-157.

Lehninger, A. L., 1975, Biochemistry, 2nd ed. Worth, New York, 1031-1056.

Lewin, R., 1982, RNA can be a catalyst. Science 218, 872-874.

Matsuno, K., 1982, A theoretical construction of protobiological synthesis: From amino acids to functional protocells. Int. J. Quantum Chem. QBS9, 181-193.

Matsuno, K., 1984, Determinism and freedom, in: Individuality and Determinism, S. W. Fox, ed., Plenum Press, New York.

Melius, P., 1979, Non-random, non-ribosomal assembly of amino acids in proteins and proteinoids. BioSystems 11, 125-132.

Melius, P., 1982, Structure of thermal polymers of amino acids. BioSystems 15, 275-280.

Melius, P., and Sheng, J. Y. P., 1975, Thermal condensation of a mixture of six amino acids. Bioorg. Chem. 4, 385-391.

Mizutani, H., and Ponnamperuma, C., 1978, The effect of polynucleotides on the dimerization of glycine, in: The Origin of Life, pp. 273-277, H. Noda, ed., Japan Sc. Soc. Press, Tokyo.

Monod, J., 1971, Chance and Necessity. Alfred A. Knopf, New York.

Morgan, T. H., 1932, The Scientific Basis of Evolution. W. W. Norton, New York.

Nakashima, T., Jungck, J. R., Fox, S. W., Lederer, E., and Das, B. C., 1977, A test for randomness in peptides isolated from a thermal polyamino acid. Int. J. Quantum Chem. QBS4, 65-72.

Przybylski, A. T., Stratten, W. P., Syren, R. M., and Fox, S. W., 1982, Membrane, action, and oscillatory potentials in simulated protocells. Naturwissenschaften 69, 561-563.

Rose, S., ed., 1982, Against Biological Determinism. Allison and Busby, London.

Schrödinger, E., 1946, What Is Life? Macmillan, New York.

Wilson, E. O., 1975, Sociobiology. Harvard University Press, Cambridge, Massachusetts.

Wilson, E. O., 1969, On Human Nature. Harvard University Press, Cambridge, Massachusetts.

Winchester, A. M., 1977, Genetics, 5th ed. Houghton Mifflin, Boston, 411.

DETERMINISM AND FREEDOM IN EARLY EVOLUTION

Koichiro Matsuno

Laboratory of the National Foundation of Cancer Research
at the Institute for Molecular and Cellular Evolution
University of Miami, Coral Gables, Florida 33134
Technological University of Nagaoka, Nagaoka 949-54, Japan

INTRODUCTION: PROTOBIOLOGICAL AND BIOLOGICAL INFORMATION

Where do protobiological and biological information originate,
in the initial conditions, in the external boundary conditions, or
within matter itself? This chapter is an attempt to answer the
question on physical grounds.

The present formulation of the problem necessarily addresses
itself to how information is related to physical processes (Fox and
Matsuno, 1983). Information, by its original definition, is an
unambiguous measure of ambiguity on the part of the observer (Shannon and Weaver, 1949). However, protobiological and biological
information refer to the measure of ambiguity of indefiniteness on
the part of matter itself, not to that on the part of the observer.
The indefiniteness of matter is in the internal freedom which matter
allows in itself. The informational aspect of physical processes is
thus in how the number of degrees of internal freedom of a material
system are constrained either by endogenous or by exogenous means.

Physical investigation of the origin and evolution of biomolecules always goes along with the information to identify how these
biomolecules are internally constrained. This question, in principle, can be answered by identifying the initial conditions of
matter, the external boundary conditions, and the law of transformation that specifies how matter evolves during development. The
initial conditions refer to the time-point information, and the
external boundary conditions refer to the information of exogenous
origin. The law of transformation refers to the information of
dynamic process or the information of how a later state relates to
an earlier state, whereas the time-point information includes no

dynamics and no transformation. Thus, a more specific question
we shall consider is: Given a time-point information of matter at
a certain initial time, can the information of dynamic processes be
constrained with them? If so, how?

The present question portends at least two possible answers.
The first one is to admit that the information of exogenous origin
dominates the content of the information of dynamic process. This
is tantamount to saying that the origin and evolution of biomolecules is due principally to external agents that are free from any
conceivable constraints. The Neo-Darwinian view of evolution
(Eden, 1967; Ho and Saunders, 1979) is of just this sort. The
second alternative is exactly opposite to the first and emphasizes
that the law of endogenous transformation dominates the constraint
upon the information of dynamic process (Fox, 1983; Matsuno, 1983).
This alternative view observes that evolution of matter is endogenous and thus turns out to be internally determinate (Fox and
Nakashima, 1983) in the sense of being free from unconstrained
external agents.

We shall discuss the physical process of how the information
of dynamic processes is constrained with time, as making explicit
the relationship among the initial conditions, the external boundary
conditions and the law of transformation. This analysis will confirm on the physical basis the second view that evolution of matter
is internally determinate.

SECOND LAW OF THERMODYNAMICS

One of the principles which relate the time-point information
to the law of transformation, or the information of dynamic process,
is the second law of thermodynamics. Informationally, the second
law of thermodynamics states that the information about the initial
conditions of a system that is not in thermodynamic equilibrium
with the surroundings will finally be lost from the system itself.
The system cannot restore the lost information later by itself,
without disturbing its surroundings. The irrevocable loss of the
time-point information about an earlier state gives the information
of dynamic process an irreversible character.

An origin of the irreversibility associated with the information of dynamic processes is Boltzmann's Stosszahlansatz; i.e.,
the principle of molecular chaos, which serves as a foundation of
statistical mechanics. An interacting many-particle system, if the
principle of molecular chaos is applicable, fails to return to its
earlier state due to the increase in both the value range of physical variables and the number of degrees of internal freedom. The
principle itself says that particles confined in a region with a
small solid angle at a certain time will be found in the region
with a larger solid angle at a succeeding instant. The information

of a dynamic process thus becomes less constrained with time.
Because of this loosening of internal constraints, however, the
principle of molecular chaos cannot work as the law of transformation underlying the origin and evolution of biological systems.
The principle of molecular chaos is opposite to the principle of
material self-assembly.

One more conceivable mechanism for irreversibility is a monotonous decrease in the number of degrees of internal freedom. What
is required for reasoning a monotonous decrease in the number of
degrees of internal freedom is to find a process that can constrain
the degrees of internal freedom successively. In fact, one can
verify (Matsuno, 1982, 1983), as we shall visualize, that a material
aggregate open to material flow constrains the number of degrees of
internal freedom. Biological systems are certainly open material
aggregates.

In order to visualize the physical origin of constraints, let
us imagine that a material aggregate open to material flow changes
its interaction at a certain interface with the exterior so as to
maintain the continuity of material flow at the interface. The
effect of interaction change propagates inside the material aggregate and induces new causes of the further interaction changes for
restoring the continuity of material flow at other interfaces with
the exterior. The spillover of interaction changes for the continuity of material flow, or what we call material flow equilibration
is always the preceding equilibration. As a result, the multiplicative constraints upon open material aggregates yield a monotonous
decrease in the number of degrees of internal freedom because each
operation of material flow equilibration provides an internal constraint with which the constituent molecular elements must comply
altogether. The information of the dynamic process due to material
flow equilibration turns out to be increasingly constrained with
time, contrary to the case of the principle of molecular chaos that
induces the loosening of the internal constraint.

We have considered two modes of irreversibility. One is the
monotonous loosening of constraint with time due to the principle
of molecular chaos. The other is the monotonous enhancement of
constraint due to material flow equilibration of open material
aggregates. As far as open material aggregates are available, the
irreversibility of the second law of thermodynamics is characterized by the information of the dynamic process constrained successively with time.

PERPETUAL DISEQUILIBRIUM

The monotonous enhancement of constraint upon an open material
aggregate is thus found to be a candidate for the law of

transformation that underlies the origin and evolution of biological
systems, because the associated information of dynamic process is
monotonously constrained with time, as these biological systems
suggest. We shall investigate more details of the endogenous process of enhancing the constraints which are multiplicative with time.

Any open material aggregate has its capacity of recognizing its
exterior through material interaction. To say that matter can
recognize its exterior implies simply that it has the capability of
interacting with the exterior. A consequence of recognizing the
exterior is the process of how the interaction with the exterior
could change or not change accordingly. The internal molecular
configuration of an open material aggregate interacting with its
exterior materializes through the recognition of its exterior followed by the renewed implementation of the interaction configuration. Here, the interaction configuration is used as synonymous
with the spatial disposition of interacting molecules. What is
required in this renewal process of interaction configuration is
the action for fulfilling the continuity of material flow or the
equilibration of material flow.

Every interacting atom and molecule in an open material aggregate is an agent of recognizing and acting upon other atoms and
molecules. This is due to the fact that the interaction change
propagates in the medium always with a finite velocity. The recognition on the part of an arbitrary interacting element is the
process of how it senses the way of interaction with others, and
the action is the process of how it adjusts the interaction with
others. Internal structural identification as a process of recognizing others is followed by internal structural realization
as a process of acting upon others. If we call the difference
between internal structural identification and realization as disequilibrium, an open material aggregate is found to be in perpetual
disequilibrium. This is because the action for fulfilling the continuity of material flow spills over in the neighborhood successively and indefinitely inside the aggregate. An external observer
is not allowed to identify the structure of interaction configuration in the presence of disequilibrium (Minsky, 1980; von Neumann,
1966) since the structure is in the process of changing itself in
an unprecedented manner. Only in the absence of disequilibrium,
or in equilibrium, can one identify the structure of interaction
configuration externally because this structure lacks the internal
mechanism of changing itself. The structure of interaction configuration in equilibrium, however, deprives itself of the capability of producing evolutionary novelties.

Perpetual disequilibrium between internal structural identification and realization in a system of interacting and reacting
chemical species prohibits us from writing the equation of motion,
or the kinetic equation (Nicolis & Prigogine, 1977), that relates

arbitrary state variables, such as the concentrations of chemical
species, to their temporal variation in an unambiguous manner (von
Hayek, 1978). The equation of motion requires a simultaneous identification of both state variables and their temporal variations
by an external observer. However, the measurements of both state
variables and their temporal variations cannot proceed in a simultaneous manner (Atkins, 1982; Pattee, 1977, 1979), since one cannot
claim a precise measurement of state variables while their values
are varying. One cannot also claim a precise measurement of variations in state variables while maintaining the values of state
variables to be measured precisely. Therefore, the equation of
motion as a simultaneous expression of both state variables and
their temporal variations does require a specific structural constraint that can relate these complementary quantities unambiguously
and that by itself remains invariant. In fact, if the disequilibrium between internal structural identification and realization
persists as it always does in open material aggregates, one cannot
expect such an invariant structural constraint that could establish
the equation of motion (von Hayek, 1978; Matsuno, 1980).

The present consideration now raises a serious question of how
to integrate both the perpetual disequilibrium character of evolving matter and the equilibrium character of the Schrödinger equation
as a quantum-mechanical equation of motion. The motivation lying
behind this question is the recognition that the Schrödinger equation must be observed as far as the evolution of matter is recognized as a quantum-mechanical process. We shall, however, find
that there is no serious difficulty in the integration. The invariant equilibrium character of configuration which the Schrödinger
equation implements is stochastic in its nature. What is invariant
is an ensemble of structures of interaction configuration. The
quantum-mechanical equation of motion does not specify which structure of interaction configuration would materialize among the
members constituting the ensemble. Evolution of open material
aggregates, on the other hand, makes it a crucial problem to identify which structure of interaction configuration materializes in
reality and how it changes with time. The quantum-mechanical equation of motion does not exclude the possibility that a particular
structure of interaction configuration changes itself with time
endogenously. The only requirement upon the equation is that the
ensemble of structures of interaction configuration remains
unchanged regardless of whether the ensemble is directly observable.

In particular, the number of possible members constituting
the ensemble would be far greater than the number of realizable
ones as far as the evolutionary process is concerned. The stochastic prediction based upon the quantum-mechanical equation of
motion can thus tell us that the evolutionarily significant trajec-
tory of change in the structure of interaction configuration would
maintain its probability measure at almost zero. This observation,

however, by no means implies that the evolutionary process is a
sequence of events that are extremely unlikely. This vanishingly
small probability measure of the evolutionary process is nothing
more than its relative frequency measured against a hypothetical
ensemble of structures of interaction configuration.

Evolution of open material aggregates requires a form of
quantum mechanics beyond its widely-quoted stochastic interpretation (Jammer, 1974; d'Espagnat, 1976; Pagels, 1982). What makes
the quantum mechanics of evolutionary processes go beyond the
standard stochastic interpretation is the law of transformation
originating in the perpetual disequilibrium between internal structural identification and realization on the part of open material
aggregates.

DIVERSITY OF DISTINGUISHABLE INDIVIDUALS

We shall try to figure out some of the qualitative aspects of
the quantum-mechanical process in perpetual disequilibrium. For
this, let us consider a primordial soup (Florkin, 1975) as a system
of interacting and reacting chemical species. A principal quantum-
mechanical character of the primordial soup is the immense multitude of individual molecules indistinguishable from each other;
e.g., the prodigious number of amino acid molecules of the same
type, and the extremely limited number of types of distinguishable
individual molecules; e.g., the limited number of different types
of amino acid. Polymers and protocells (Fox and Dose, 1977) formed
in the primordial soup serve as open material aggregates. The
internal constraint due to material flow equilibration of open
material aggregates always imposes a new internal regulation to be
maintained among the then existing chemical species. The internal
regulations of this sort accumulate in the primordial soup with
time. A consequence of such a multiplicative internal regulation
among the existing chemical species is the construction of a new
species from them that differs from any of the pre-existing species. The newly formed species is different from those expected
from a random aggregation of the pre-existing species, since the
process of internal regulation originating in material flow equilibration is self-constrained and self-limited as noted already.
We thus observe that the number of types of distinguishable individual species found in the primordial soup can increase with time
at the expense of the decrease in the number of indistinguishable
individuals due to the quantum-mechanical process in perpetual
disequilibrium.

Experimental evidence of forming new chemical species from
the pre-existing ones is available from thermal polymerization of
amino acids in a simulated primordial soup (Fox and Harada, 1958).
In fact, thermal proteins as new species constructed from the

pre-existing amino acids are already self-constrained and self-
limited about their amino acid sequences compared with a merely
hypothetical random polymerization (Nakashima et al., 1977; Hartmann et al., 1981). The number of different major types of thermal
proteins, on the one hand, is quite limited, at most three to six,
if the initial amino acid composition ratios are fixed (Fox and
Nakashima, 1967); Melius and Sheng, 1975). On the other hand,
however, if the initial composition ratios are varied (Fox and
Waehneldt, 1968), the additive number of different types of thermal
proteins would increase accordingly.

During thermal polymerization of amino acids, the growing
polymer as an open material aggregate successively changes its
internal structure through indefinite spillovers of interaction
change. Each amino acid molecule fixed in a sequence of thermal
protein is determined in a consistent manner with all the other
amino acids participating in the sequence. Each amino acid adjusts
itself in relation to all the other amino acids found in the same
polymer. This implies that an amino acid in isolation intrinsically carries an internal freedom that could be constrained during
thermal polymerization in accordance with the internal regulation
of an open material aggregate. The physical source of a new different species; e.g., thermal protein, is found in the indefiniteness carried by the pre-existing species, e.g., amino acids. The
internal regulation imposed upon the previous indefiniteness is the
physical process underlying the construction of the new species.
The indefiniteness inherent in amino acids is transferred to
thermal proteins to a lesser extent because of the cumulative internal regulation. The indefiniteness transferred to thermal proteins
can serve as a source of constructing further different species.

Phase-separation of proteinoid microspheres from an aqueous
solution of thermal proteins (Fox and Dose, 1977) is an experimental demonstration of further constraints acting upon the indefiniteness carried by thermal proteins. Thermal proteins in proteinoid
microspheres are more constrained than thermal protein molecules in
isolation. What is more, proteinoid microspheres maintain their
own indefiniteness as transferred from thermal proteins, although
to a lesser extent. It is this indefiniteness associated with proteinoid microspheres that makes themselves later reproductive at
least heterotrophically (Fox et al., 1967), since the reproduction
of proteinoid microspheres is a consequence of applying additional
constraints upon the parental microspheres.

What characterizes the evolution of matter as a quantum-
mechanical process in perpetual disequilibrium is the increase in
the number of types of distinguishable individual species at the
expense of the decrease in the number of indistinguishable individuals. Emergence of new distinguishable individuals is an internally regulated physical process of open material aggregates and

is never a chance event. An evolutionarily new species finds its roots not in the information of exogenous origin, but in the information of dynamic process specifying how the quantum-mechanical process in perpetual disequilibrium proceeds endogenously.

LAWS: STATISTICAL AND EVOLUTIONARY

Given an interacting many-molecule system, the structure of interaction configuration among those molecules would remain fixed if the disequilibrium between internal structural identification and realization disappears. However, the very fact of terrestrial evolution is witness to a form of perpetual disequilibrium. This observation raises such a question of how the process of perpetual disequilibrium could be reconciled with the physical laws which find their basis upon repeatedly-tested synchronic processes irrespective of their historical contexts.

If the primordial soup is the case under consideration, one can recognize the immense multitude of molecules indistinguishable from each other, in contrast to the extremely limited number of different types of molecule. Furthermore, if the endogenous process of generating distinguishable individuals can be neglected, it is possible to single out a set of statistical laws by averaging over an immense ensemble of individual species. Invariant statistical laws (Schrödinger, 1946) seek their foundation upon the invariance of the ensemble of individual species, and thus make themselves synchronic laws (Van Kampen, 1981; Smith and Morowitz, 1982). The evolutionary character is ruled out from the invariant statistical or stochastic ensemble.

If one tries to find a basis of synchronic physical laws in the invariant structural property (Wigner, 1964) such as an invariant statistical ensemble, the evolutionary process allowing the generation of new distinguishable individuals could be deemed outside of the realm specified by those physical laws. The evolutionary process continues to change the very ensemble of individual species through the never-ending generation of new distinguishable species. However, it is also important to note that the presence of an invariant structural property is not the only source of synchronic physical laws. The functional process underlying a realization of physically invariant principles (e.g., Lahiri, 1977) can also provide a basis of synchronic physical laws, since the testable physically invariant principle is always accompanied by the process realizing itself. The principle of material conservation, for instance, is accompanied by the process of fulfilling the continuity of material flow. This functional invariance is more than synchronic. It is also chronic in the sense that the internal process of fulfilling the continuity of material flow persists all through the time development.

Material flow equilibration of open material aggregates is founded upon the functional invariance underlying the process of fulfilling the continuity of material flow. The functional invariance can maintain the perpetual disequilibrium between internal structural identification and realization. This observation makes it evident that material flow equilibration is an evolutionary law, in addition to being a repeatedly testable physical law, because of its chronic character of invariance. The range of validity of invariant statistical laws is limited to the region where an invariant statistical ensemble is available. In contrast, the evolutionary law finds its basis upon the functional invariance which remains valid irrespective of whatever ensemble of individual species may be available in reality.

DETERMINISM AND FREEDOM

In the absence of disequilibrium between internal structural identification and realization, an interacting many-element system is found to be in equilibrium in the sense that the structure of interaction configuration among the constituent elements lacks any tendency of changing itself. Since the structure of interaction configuration in equilibrium remains invariant with time, the later time development of such a system is completely specified and determined by the initial condition. The determinism based upon the structure of interaction configuration in equilibrium, or simply determinism in equilibrium, makes it possible to determine the past from the data of the future as well as the future from the data of the past. A well-known example of determinism in equilibrium is classical mechanics. The centrifugal interaction configuration of classical bodies remains invariant. The initial condition is simply irrevocable. There is no room for making choices internally during time development. In other words, freedom does not coexist in the scheme of determinism in equilibrium.

Determinism in equilibrium, as Laplace perceived it (e.g., Pattee, 1979), would be possible only if an isolated system such as celestial mechanics considers could be available. In reality, however, any interacting many-element system inevitably interacts with its exterior. The structure of interaction configuration that can remain indefinitely in equilibrium is not available. One typical view for overcoming this difficulty is the hypothesis that the structure of interaction configuration in equilibrium may sometimes be disturbed by unconstrained external agents. This view, when applied to evolutionary processes, turns out to be Neo-Darwinian (Eden, 1967; Maze and Bradfield, 1982). Evolutionary novelties are always sought in unconstrained external agents. The structure of interaction configuration temporarily in equilibrium provides a deterministic character and its punctuation by intermittent disturbances is proposed to symbolize the aspect of freedom

in evolution (Monod, 1971). The structure of interaction configuration temporarily in equilibrium, or simply determinism temporarily
in equilibrium, punctuated by unconstrained external disturbances,
is more realistic than determinism in equilibrium that does not
allow any interfering external disturbances. However, the question
that must be answered is whether the structure of interaction configuration temporarily in equilibrium can be justifiable.

    Any biological system is open to material flow. We have
already observed that any open material aggregate maintains its
structure of interaction configuration in perpetual disequilibrium
because of the persistent disequilibrium between internal structural identification and realization. The internal action for fulfilling the continuity of material flow at one place never fails
to cause the analogous action at others in a successive manner.
The action for fulfilling the continuity of material flow, or material flow equilibration, always provides a constraint with which
all the constituent material elements must comply. The indefiniteness carried by these material elements is thus successively constrained. The process of successive constraints is internally
determinate because of the explicit physical link between the cause
and the effect of these constraints. On the other hand, what is
carried by the internally determinate process is the indefiniteness
about the structure of interaction configuration, although constrained successively with time. The evolution of open material
aggregates is deterministic in the sense that the evolutionary
process is endogenously regulated, but it carries the indefiniteness which allows matter to be in perpetual disequilibrium between
internal structural identification and realization. The outcome
is determinism in disequilibrium. The process of constraining the
indefiniteness of matter with regard to its interaction configuration is internally determinate.

    If we associate the connotation of freedom with the term
indefiniteness, the indefiniteness carried by determinism in disequilibrium would yield a constrained freedom. This differs from
the unconstrained freedom envisaged in Neo-Darwinian evolutionary
process. The structure of interaction configuration temporarily in
equilibrium punctuated by unconstrained variations is not justifiable in the real evolutionary process, since the interaction configuration in perpetual disequilibrium is common to any biological
system. This does not, however, exclude the possibility that an
external agent may intervene in the evolutionary process, for
instance, through an impact of asteroid on terrestrial evolutionary
process. What does matter is that the interaction configuration
of material aggregates involved in the evolutionary process is in
perpetual disequilibrium.

    Determinism temporarily in equilibrium punctuated by uncon-
strained disturbances obtains its evolutionary novelties from

these unconstrained disturbances. Because of this character, Neo-Darwinian evolution acquires its evolutionary novelties from external agents. The structure of interaction configuration temporarily in equilibrium serves only for the purpose of preserving the once acquired novelties. The determinate aspect of Neo-Darwinian evolution refers only to the preservation of the evolutionary novelties originating in unconstrained external agents. In other words, Neo-Darwinian evolution is totally indeterministic with regard to its novelties. This is consonant with the view that freedom in the Neo-Darwinian view is free from any constraints and remains absolute (Monod, 1971). The determinate character of evolution and the origin of freedom perceived in the Neo-Darwinian scheme are thus completely separated from each other.

Determinism in disequilibrium, on the other hand, observes that the origin of evolutionary novelties is endogenous and internal. One thus recognizes (Dobzhansky, 1981) that evolutionary novelties are already latent in the pre-existing material aggregates and that only the lapse of time is required for transforming those latent into the actual novelties to be observed. The physical source of evolutionary novelties is the indefiniteness carried by the evolutionary precursors. Freedom survives only in the constrained form as far as the carried indefiniteness is not exhaustively spelled out. A consequence of the endogenous constraint upon the freedom carried by the evolutionary precursors is the decrease in the number of individuals indistinguishable from each other and the emergence of new distinguishable individuals.

Determinism in equilibrium, e.g., classical mechanics, has no room for freedom, in addition to being irrelevant to the evolutionary process in the real world because of its hypothetical isolatedness. In contrast, determinism temporarily in equilibrium punctuated by external disturbances admits absolute freedom, especially in the genesis of evolutionary novelties. Between these two extremes, one can find the place for determinism in disequilibrium that underlies the real evolutionary processes. The freedom allowed in determinism in disequilibrium continues to be constrained through the internally determinate process. Fundamental to the freedom consistent with determinism in disequilibrium is the recognition that any interacting element, whether physical, chemical, or biological, maintains in itself a certain indefiniteness which can never be exhaustively identified in its isolation.

REFERENCES

Atkins, P. W., 1982, Quantum theory: Certainty in uncertainty. A book review on "The Cosmic Code: Quantum Physics as the Language of Nature," by Heinz R. Pagels, Nature 296, 503.

d'Espagnat, B., 1976, Conceptual Foundations of Quantum Mechanics.
    W. A. Benjamin, Reading, Massachusetts.
Dobzhansky, T., 1981, From potentiality to realization in evolution.
    Anthro. UCLA 7, 71–82.
Eden, M., 1967, Inadequacies of Neo-Darwinian evolution as a sci-
    entific theory, in: Mathematical Challenges of Neo-Darwinian
    Interpretation of Evolution, P. S. Moorehead and M. M. Kaplan,
    eds., Wistar Institute Press, Philadelphia, pp. 5–12.
Florkin, M., 1975, Ideas and experiments in the field of prebiologi-
    cal chemistry, in: Comprehensive Biochemistry, Vol. 29B,
    M. Florkin and E. H. Stotz, eds., Elsevier, Amsterdam, pp. 231–
    260.
Fox, S. W., 1983, Proteinoid experiments and evolutionary theory,
    in: Beyond Neo-Darwinism, M. W. Ho and P. T. Saunders, eds.,
    Academic Press, London.
Fox, S. W., and Harada, K., 1958, Thermal copolymerization of amino
    acids to a product resembling protein. Science 128, 1214.
Fox, S. W., McCauley, R. J., and Wood, A., 1967, A model of primi-
    tive heterotrophic proliferation. Comp. Biochem. Physiol. 20,
    773–778.
Fox, S. W., and Nakashima, T., 1967, Fractionation and characteriza-
    tion of an amidated thermal 1:1:1 proteinoid. Biochim.
    Biophys. Acta 140, 155–167.
Fox, S. W., and Waehneldt, T. V., 1968, The thermal synthesis of
    neutral and basic proteinoids. Biochim. Biophys. Acta 160,
    246–249.
Fox, S. W., and Dose, K., 1977, Molecular Evolution and the Origin
    of Life, rev. ed. Marcel Dekker, New York.
Fox, S. W., and Nakashima, T., 1983, Endogenously determined varia-
    tions as substrates for natural selection, in: Individuality
    and Determinism, S. W. Fox, ed., Plenum Press, New York.
Fox, S. W., and Matsuno, K., 1983, Self-organization of the proto-
    cell was a forward process. J. Theor. Biol. 101, 321–323.
Hartmann, J., Brand, M. C., and Dose, K., 1981, Formation of spe-
    cific amino acid sequences during thermal polymerization of
    amino acids. BioSystems 13, 141–147.
Ho, M. W., and Saunders, P. T., 1979, Beyond Neo-Darwinism: An
    epigenetic approach to evolution. J. Theor. Biol. 78, 573–591.
Jammer, M., 1974, The Philosophy of Quantum Mechanics. Wiley, New
    York.
Lahiri, A., 1977, The functional approach in biology. BioSystems
    9, 57–68.
Matsuno, K., 1980, Disequilibrium dynamics of autonomous systems
    and their structural transformation. Int. J. Gen. Syst. 6,
    75–82.
Matsuno, K., 1982, A theoretical construction of protobiological
    synthesis: From amino acids to functional protocells. Int.
    J. Quant. Chem., QBS9, 181–193.
Matsuno, K., 1984, Open systems and the origin of protoreproductive
    units, in: Beyond Neo-Darwinism, M. W. Ho and P. T. Saunders,
    eds., Academic Press, London.

Maze, J., and Bradfield, G. E., 1982, Neo-Darwinian evolution: Panacea or Popgun. Syst. Zool. 31, 92–95.

Melius, P., and Sheng, J. Y-P, 1975, Thermal condensation of a mixture of six amino acids. Bioorg. Chem. 4, 385–391.

Minsky, M., 1980, Commentary on J. R. Searle, minds, brains and programs. Behavior and Brain Sciences 3, 417–457.

Monod, J., 1981, Chance and Necessity. Alfred A. Knopf, New York.

Morgan, T. H., 1932, The Scientific Basis of Evolution. W. W. Norton and Company, New York.

Nakashima, T., Jungck, J. R., Fox, S. W., Lederer, E., and Das, B. C., 1977, A test for randomness in peptides isolated from a thermal polyamino acid. Int. J. Quant. Chem. QBS4, 65–72.

Nicolis, G., and Prigogine, I., 1977, Self-Organization in Nonequilibrium Systems. Wiley, New York.

Pagels, H. R., 1982, The Cosmic Code: Quantum Physics as the Language of Nature. Simon & Schuster, New York.

Pattee, H. H., 1977, Dynamic and linguistic complementarity in complex systems. Int. J. Gen. Syst. 3, 259–266.

Pattee, H. H., 1979, The complementarity principle and the origin of macromolecular information. BioSystems 11, 217–226.

Shannon, C. E., and Weaver, W., 1949, The Mathematical Theory of Communication. University Of Illinois Press, Urbana.

Schrödinger, E., 1945, What Is Life? Cambridge University Press, London.

Smith, T. F., and Morowitz, H. J., 1982, Between history and physics. J. Mol. Evol. 28, 265–282.

Van Kampen, N. G., 1981, Stochastic Processes in Physics and Chemistry. North-Holland, New York.

von Hayek, F. A., 1978, New Studies in Philosophy, Politics, Economics, and the History of Ideas. Routledge & Kegan Paul, London, pp. 23–34.

von Neumann, J., 1966, Theory of Self-Reproducing Automata, edited and completed by A. W. Burks. University of Illinois Press, Urbana.

Wigner, E., 1964, Events, laws of nature, and invariance principles. Science 145, 995–999.

PARTICIPANTS

THOMAS J. BOUCHARD, JR.
Psychology Department
University of Minnesota
Minneapolis, MN

J. DONALD CAPRA
Health Science Center
University of Texas
Dallas, TX

KLAUS DOSE
Institute for Biochemistry
Johannes Gutenberg University
Mainz, Germany

SIDNEY W. FOX
Institute for Molecular and
  Cellular Evolution
University of Miami
Coral Gables, FL

GARRETT HARDIN
Department of Biological
  Sciences
University of California
Santa Barbara, CA

DANIEL E. KOSHLAND, JR.
Department of Biochemistry
University of California
Berkeley, CA

WERNER R. LOEWENSTEIN
Department of Physiology and
  Biophysics
University of Miami
Miami, FL

CHARLES B. METZ
Institute for Molecular and
  Cellular Evolution
University of Miami
Coral Gables, FL

ALBERTO MONROY
Stazione Zoologica
Napoli, Italy

WILLIAM J. RUTTER
Department of Biochemistry and
  Biophysics
University of California
San Francisco, CA

HENRY KING STANFORD
Office of President Emeritus
University of Miami
Americus, GA

ETHEL TOBACH
Department of Animal Behavior
American Museum of Natural
  History
New York, NY

INDEX